ROY DENNIS, MBE is a field ornitholo
He has worked in the highlands anc
1959, most notably on the conservatio of nature
and wild places. He is a specialist in raptor conservation and reintroductions in the UK and abroad, having been involved with osprey, red kite, golden eagle and sea eagle reintroduction projects. His satellite tracking studies since 1999 have broken new ground by linking active data with website maps for public interest and enjoyment. He has long been an advocate for ecological restoration on a large scale and the reintroduction of lost mammals to Scotland, starting with beaver and lynx. In 1992, Roy was awarded an MBE for services to nature conservation in Scotland and in 2004 won the RSPB Golden Eagle Award for being the person who had done most for nature conservation in Scotland in the last 100 years. He is a writer, lecturer and broadcaster.

'Now 81, Dennis is possibly the UK's most senior and influential conservationist you may never have heard of … *Restoring the Wild* covers Dennis's career, which he built on the profound knowledge of ecology and species interaction. It is a book based on his diaries of 60 years, which reveal the sheer number of reports, town hall lectures, licences, risk assessments, acronyms, obstacles and setbacks before the excitement comes … Exhilarating'

KATHLEEN JAMIE, *New Statesman*

'If you want to find out how to help wildlife survive, thrive, and expand their own horizons, Roy Dennis is the very man: according to the RSPB, no-one else has done more for nature conservation in Scotland in the last 100 years. Informed by sixty years of fieldwork, his new book is a comprehensive guide to how well – or poorly – we are placed to rewild our skies, woods and waterways'

DAVID ROBINSON, *Books from Scotland*

'Dennis is the most significant conservationist you've probably never heard of, and possessed of a radicalism that would startle the most outspoken young environmentalist … Dennis's vision of how to halt the extinction crisis and restore lost habitats and species in Britain deserves attention because it is rooted in 60 years of pioneering conservation action … Like Dennis, the book is modest, deeply informative and profoundly hopeful, in that it shows a new generation how to bring back lost species'

PATRICK BARKHAM, *Guardian*

'Roy's passion and love of wildlife is visible in every crevice of the book … a tantalising and inspiring read' *The Scottish Field*

'A wonderful book, steeped in knowledge and experience of nature and of the more practical ends of nature conservation … No one else could write this book from personal experience – it's a treat … A joy to read, and gives plenty of information but also food for thought' MARK AVERY

'A leading figure in UK reintroductions, Roy explains the process and necessity of bringing back our extinct apex predators'

*Nature's Home*

'Inspiring stories … Tales from the frontline of conservation, offering insights into the complexities that go with reintroducing species to Britain. [Dennis] also makes the case for why wildlife management is vital for our nation's continued wellbeing'

*The Bay* magazine

'Really interesting, a passionate guide and how-to-manual for conservation in this country'

STEPHEN RUTT, author of *The Seafarers* and *Wintering*

# RESTORING THE WILD

## Rewilding Our Skies, Woods and Waterways

ROY DENNIS

WILLIAM
COLLINS

William Collins
An imprint of HarperCollins*Publishers*
1 London Bridge Street
London SE1 9GF

WilliamCollinsBooks.com

HarperCollins*Publishers*
1st Floor, Watermarque Building, Ringsend Road
Dublin 4, Ireland

First published in Great Britain by William Collins in 2021
This William Collins paperback edition published in 2022

2023 2025 2024 2022
2 4 6 8 10 9 7 5 3 1



A catalogue record for this book is
available from the British Library

ISBN 978-0-00-836882-1



Typeset in Dante MT Std
Printed and bound in the UK using 100%
renewable electricity at CPI Group (UK) Ltd

MIX
Paper from
responsible sources
FSC™ C007454

This book is produced from independently certified FSC™ paper
to ensure responsible forest management.

For more information visit: www.harpercollins.co.uk/green

Dedicated, with love, to my children
Rona, Gavin, Roddy and Phoebe

# CONTENTS

Author aged 11 holding tame young jackdaw and shelduck at his home in Hampshire.

# INTRODUCTION

I've been interested in helping wildlife since I was a young boy in the early 1950s. Then, it was a matter of rearing young birds and animals that I found abandoned, even though some of them may not actually have needed my help. Jackdaws were a favourite of mine, and I have happy memories of rearing three shelducklings which I found wandering along the road in coastal Hampshire. What a thrill it was to see them fly after me, a squadron of outriders as I rode my bike. Little did I know then that this calling would form the basis of my life's work.

It has always been my view that our aim as nature conservationists should be to make rare species more common and more secure for the future. My mind was always on ecological restoration, now more often called rewilding. In an era of appalling declines in nature, leading to lost or isolated populations, mainly due to human activity, I believe that species translocations and reintroductions must become standard wildlife management.

This book is about the many species restoration or rewilding projects that I have been involved with in my life – so far. It covers recovery projects (such as goldeneye), translocations (such as red squirrel) and reintroductions (such as sea eagle) – a complex mix. I wanted to write a book about these exciting projects, making certain that I explained the failures as well as the successes, with an emphasis on the early stages. Specific dates, of course, are certain: the young sea eagles arrived at Fair Isle on 28 June 1968; the first red kites from Sweden came to the Scottish Highlands on an RAF Nimrod on 20 June 1989. But much of the story is, as so often, a record preserved in the notebooks, diaries, letters and memories of the writer, and may differ from other people's accounts. I've found it fascinating to go back and read my writings and view my photos, stretching back to my childhood, often long forgotten. There are several levels of memory: my writings, especially my diaries and notebooks, give me the exact dates, places and what we were doing, but it's when I look at the photographs taken at the time that the colours, textures, scents and sounds flood back.

Some events I can no longer properly remember, but I do know that I was at the Scottish Ornithologists' Club's northern counties conference in Inverness in November 1962, when I was 22. The speakers were prominent ornithologists of the time: George Dunnet and Harry Milne of Aberdeen University, Derek Mills, Ian Pennie and Joe Eggeling. A report of the conference in the journal *Scottish Birds* notes: 'Roy concluded the morning's session with an entertaining talk on the encouragement of rare breeding species, in which he put forward some excellent ideas for attracting these visitors on a wider scale.' I cannot find a record of that talk but did find some scribblings in which I worked out that a pair of (reintroduced) sea eagles could be worth £15,000 a year in tourism revenues to northern Sutherland. Times change and years go by, and now the sea eagles are worth £5 million a year to the Isle of Mull alone.

It's hard to believe now that, when I started birding, books on birds and mammals were relatively scarce and new titles were eagerly awaited. What we did have was a marvellous series of Victorian-era books on birds and mammals, especially the county or regional faunas written at the start of the 20th century by JA Harvie-Brown and his contemporaries. On dark nights or rainy days I remember the thrill of reading the likes of *The Vertebrate Fauna of the North West Highlands and Skye*, to find vivid details of sea eagles, nowadays called white-tailed eagles, swirling above their breeding sites. I would read about places, which in the 1960s lacked that iconic tag of 'wildness', or tales of wolves and bears in James Ritchie's book *The Influence of Man on Animal Life in Scotland*, published in 1920. Who could forget the image of the people of Eddrachillis in Sutherland taking their dead to the island of Handa for burial to save the bodies from being dug up by wolves, known locally as the 'prowlers of the night'?

My most lasting memories from those old books tend to be of the incredible onslaught by man on the larger birds and mammals of the Scottish Highlands and Islands. Savage descriptions of killing the last wolf in Scotland: Mr MacQueen, who lived near Tomatin, dispatching it with his dirk (hunting knife) in a side stream of the River Findhorn, which I can now see in the distance when I drive south from my home. Or the incredible numbers of birds of prey and predatory mammals shot, trapped and poisoned by the gamekeepers creating the sporting estates for the hunting of red deer and red grouse and the fishing of salmon. On the notorious Glengarry estates near Fort Augustus, many thousands were killed. But this drive to remove competitors to the human interests of sheep farming or game sports was widespread. Some species, like the beaver, had gone much earlier, not because they were pests but because they were too valuable for their own good, their fur, medicinal oil or meat much sought after.

My generation was fortunate to be born into better times, with nature conservation starting to take a lead, and the obvious question to some of us was how we could get these species back. Most of these iconic creatures had died out not through changing climate, or loss of available food or habitat; nearly all were cases of simple, but appalling, extermination by man. And so we talked about this when we met, gossiping over a beer at a winter conference or while eating our piece on a mountain ridge. There was an exciting opportunity to redress the damage of previous generations, if we could only grasp it. What a thrilling prospect that was – but then, when young, I never foresaw how difficult it would be, and how many social and political hurdles would be thrown in our way, even by our colleagues and friends. With this book I've tried to show how some of these projects came about: the excitements, the obstacles, the failures and the successes. I've often concentrated on the early stages of projects, especially those that went on to be very successful, for those were the times when we just had to get our heads down, determined to get the job done with the help of really good friends. These projects are always teamwork.

I hope my contemporaries enjoy reading the accounts for different species, and I also hope that anyone interested in nature finds the stories fascinating and, in general, encouraging in difficult times. As always, I urge young people in school, in work, at university or freshly graduated to take note that wildlife recovery projects are very long term and need determination and perseverance. Often the social and political difficulties of starting the translocation of a species far outweigh the ecological and technical aspects. So, most importantly, anyone wishing to carry out projects must remember that a determined and positive mind is called for. It's a matter of 'when', not 'if'.

In 1995 I was invited by Professor Wolf Schröder and Christoph Promberger, large mammal specialists at Munich University, to join them on a summer expedition to the Upper Porcupine River in Canada to study bears and wolves. In Whitehorse, we met one of their friends at the wildlife division. 'Roy has a dream of restoring wolves to the Scottish Highlands,' said Christoph. His response? 'Do you work with any other species?' I told him about my projects on ospreys, eagles and red kites. 'That's good,' he replied, 'because if you work with only one species you will come to a roadblock from a person in authority, maybe even a senior colleague, and then you will have to put your project on the back burner. One day he or she will be promoted, will retire or die – and that may take years. When it happens, you must immediately push forward. So it's important to have a variety of projects moving at different speeds.' It turned out to be very pertinent advice.

Looking back now, I realise that great things have happened, but I'm disappointed. I'm disappointed that we have not had more success, especially with large mammals. The return of the red kite has been incredible, but this was because of a determined programme of rolling releases from north and south, and the strength of the long-term project team members from the RSPB*, Nature Conservancy (and its subsequent national agencies) and enthusiastic individuals and groups. Compare this to the failure, so far, to get the sea eagle breeding again in England and Wales; and the incredible opposition by farmers, politicians and even some conservationists to the return of the lynx.

Yet remember: there's always change. I think now that if we had really worked hard we could have restored beaver, lynx, crane and eagle owl to Scotland in the 1960s and '70s. There

---

* Please find a list of organisations and acronyms on page 431.

would have been opposition from some of our colleagues because, in their view, that would not have been 'natural', but in those days there was much less organised opposition from vested interests. A few really enthusiastic people with large landholdings and sufficient funds could have forged new and exciting ecological pathways. I guess our horizons and hopes were not broad enough. Since that time, the idea of the reintroduction of charismatic species has become more formalised, more difficult and attracted greater opposition, not just from parts of the land-owning, farming, forestry and fisheries communities who see their own interests threatened, but also from individuals and bodies within nature conservation. Guidelines which were intended to help became ever more restrictive and bureaucratic. One needs to learn, often by painful processes, that the early stages of a bold idea will attract the attention of doubters and opponents to change, as well as supporters. Once a project gets started there is a shift and opposition lessens, and finally – when it is successful – the early naysayers claim ownership. But that matters little to you if your principal aim is to restore nature.

Some of the projects described in this book have been written up in the scientific literature, often by others, but that will always have a limited readership. I have always been happy to give radio and television interviews at any time, and ready to help journalists and photographers to write good stories. That can have real impact but often leaves little permanent record. Our projects often featured in my illustrated lectures and talks to RSPB member groups and bird clubs, as well as to audiences in the wider community. I enjoyed doing these lectures, seeing the wide variation in numbers and interests, ranging from the evening with just one person, when the school janitor in the Uists forgot to put up the posters, to an audience of more than two thousand scientists at a major conference in Glasgow. This contact with the public at home and abroad often had a beneficial impact and I particularly like to encourage the young to push for change, bypassing the strictures of their seniors.

Finally, it's essential to realise that in the same half-century, the impacts of modern agriculture, forestry, industry, chemical pollution and humanity on the natural world have become ever more destructive. We have now reached another era, when the prospect of climate breakdown and biodiversity loss threatens life on earth. This is especially pertinent to our young people, so – alongside major ecological restoration and rewilding – there is much greater interest in and support for large-scale restoration and recovery of our native fauna and flora. It's time to get on with the task and listen to the concerns of the young, so let's look at the projects and species in detail to help them shape a better future.

White-tailed eagle. (*Photo by Mike Crutch/A9Birds*)

1

# SERENDIPITY

## *The return of the majestic sea eagle*

So much in life depends on good luck, on being in the right place at the right time, and for me, in 1967, the right place was the famous Fair Isle Bird Observatory in the Shetland Islands, where I was warden. My boss was the charismatic Scottish ornithologist George Waterston, who had set up the observatory and was the director of the RSPB in Scotland. He phoned me one day from his home at the Scottish Centre for Ornithology in Edinburgh, where he lived with his wife Irene, and told me that he wanted to reintroduce four young Norwegian sea eagles to Fair Isle. Would I rear and release them, he asked, and start by sounding out the crofters and letting him know what they thought? I was bowled over with excitement.

A decade earlier, I had been a teenage birdwatcher living in Hampshire, where one of my favourite places was St Catherine's Point, a chalk headland on the Isle of Wight reaching out into the English Channel and ideal for studying bird migration by day and night. To get there, I would catch the ferry from Portsmouth to Ryde, then take a small train before hitch-hiking the rest of the way, in those days my quickest route to out-of-the-way birdwatching sites. One day, my diary reminds me, I must have been dreaming about sea eagles – now called white-tailed eagles to distinguish them from other sea eagle species around the world – and decided that I would first of all visit Culver Cliff near Sandown. In 1780 this had been their final breeding site along the south coast of England. I walked from Bembridge, carefully choosing low tide so that I could get round the base of the cliffs and look up at the chalky white headland where the old books on Hampshire birds recorded their nest. Of course, there were no sea eagles to see that day, but I could vividly imagine them circling above me, their huge wings, bright yellow bills and white tails thrown into sharp focus by the clear, blue sky.

The white-tailed eagle breeds throughout northern countries from Greenland and Iceland, east from Scandinavia through Russia to the northern Japanese island of Hokkaido. Originally the breeding range extended south to the Mediterranean Sea, with the last nesting records on the island of Corsica in the 1950s. Some birds from northern populations migrate south in winter. These huge eagles build big, bulky nests in cliffs, in large trees and on the ground, and lay one to three eggs, rearing up to three young. They are a generalist raptor, eating a wide range of carrion and live prey including mammals, birds and fish. Seen by many as a threat to their livestock, the species had been subject to much human persecution over the centuries, resulting in their loss as a breeding species in all countries in southern Europe, with the last breeding pair in the UK on the Isle of Skye in 1916. Even so, the bird has had ancient associations with man: even Neanderthals are thought to have decorated the wing bones, while in Neolithic times people in the Orkney Islands placed the remains of sea eagles beside human remains. This was richly brought home to me when I visited the Tomb of the

Eagles on South Ronaldsay. A farmer there had discovered an underground tomb, accessible nowadays via a tunnel. Visitors can lie on a trolley and pull themselves into an underground chamber. In the last 50 years, white-tailed eagle numbers have started to increase and the bird has reclaimed lost range, such as in northeast France and the Netherlands, while populations have been reintroduced in Scotland and Ireland.

In 1958, when I was just 18, I was the assistant warden at Lundy Island Bird Observatory, where sea eagles had nested at least two centuries before, and the following year I worked as a field ornithologist at the famous Fair Isle Bird Observatory in Shetland, where I was later to become warden. On that island there had once been two pairs of 'ernes', the lovely Scottish name for sea eagles. The present-day Fair Isle place name 'Erne's Brae' is a memento of the days when they used to nest there, although they last bred on Fair Isle before 1840. At that point, of course, I had no idea that my life was to become so closely involved with sea eagles.

George Waterston's vision of restoring sea eagles to Scotland had been with him for many years; in 1959 he had helped his cousin, Pat Sandeman, when he arranged to save three sea eagles from being killed in Norway when they were still regarded as pests. Pat released them in Glen Etive in Argyllshire, but the adult was caught at a chicken farm and taken to Edinburgh, while one of the young was killed nearby in a fox trap and the third simply disappeared. Although the attempt was unsuccessful, neither man gave up on their idea, and in 1962 George joined the Norwegian expert, Johan Willgohs, when he was surveying sea eagles on the coast north of Bergen. This was a great stimulus to George, and Mike Everett, an RSPB colleague, told me that, from then on, his office walls were papered with even more photographs of sea eagles. Whenever Pat called by, the two men would enthusiastically discuss how to make another attempt. Things would quieten down, another year would go by, but in July 1966, in Cambridge, George gave a paper to the 14th World Conference of the International Committee for Bird Protection, and advocated reintroductions, especially that of sea eagles. He received support from some ornithologists but a cold shoulder from others. Foreign support was probably the turning point, with Professor Karl Voous of Amsterdam, who supported the proposal in a letter to Peter Conder, head of the RSPB, because the species was in such great decline, remarking, 'Indeed, the enthusiasm for reintroduction is a kind of contagious disease.' Kai Curry-Lindahl, the eminent Swedish ornithologist, was very supportive and gave practical advice, while the expert himself, Johan Willgohs, supported it and offered to collect four young eagles.

George discussed the idea with the National Trust for Scotland, the owners of Fair Isle, and secured their agreement. He wrote to me, asking me to go to the islanders and ask if they could arrange a meeting of the island committee in mid-September, when he would be briefly on the island. He would give a talk on sea eagles and, being George, he also asked if there could be a Fair Isle dance afterwards, which always offered a great chance to catch up on island news. The meeting took place, there was a wide-ranging discussion, and George's request did not fall on deaf ears, with the islanders giving their tacit support, with just one proviso: 'If they start eating our sheep, they'll have to go!' It would be my job to rear and release them. I was thrilled to get such a fantastic opportunity, which would turn out to add excitement to my life for many years to come.

In my archives I found a letter I wrote to George on 12 November 1967, along with a six-page memo, entitled 'Reintroduction of the white-tailed eagle to Fair Isle as a scientific experiment'. I outlined my thoughts on the project, the reasons for the birds' extinction and the potential for success. The last pair bred on Sheep Rock between 1825 and 1840. This was at a time when the island population was rising rapidly, going from 160 people in 1801 to 232 in 1841; this was also a time when sheep were becoming more common in Scotland. I said the species was most unlikely to recolonise naturally because of its sedentary nature and low numbers, so reintroduction was essential. Food on Fair Isle was plentiful and the latest seabird counts (pairs) in 1966 gave fulmar 5,000 (there were none when the sea eagle last nested), shag 1,100, eider 150, oystercatcher 88, herring gull 300, kittiwake 7,900, guillemot 5,600, razorbill 1,000, puffin 15–16,000 and black guillemot 150–200. I included a list of when they were available as prey. Fish around the isle were plentiful, with larger fish, like ling, washed up on the beaches after storms. Rabbits were very plentiful, too, with possibly 3,000 present in autumn. We needed to keep a close watch on lambs in spring. The sea eagles could increase bird tourism and the resulting income, and Fair Isle's isolation would deter future egg collectors and indiscriminate birdwatchers and photographers.

The next months were hectic for George as he drew up a proposal. Peter Conder wrote to George on 6 February 1968, explaining the need for a concise report, preferably one sheet of foolscap, to be presented to the RSPB conservation committee on 28 February. It then would have to go to the International Council for Bird Preservation, British section AGM on 6 March, to the Advisory Committee for the Protection of Birds for Scotland on 7 March, to the RSPB Council on 13 March and to the General Committee of the Nature Conservancy on 2 May. Again, George received knockbacks from various members of the ornithological establishment, but he also received support from senior scientists, including Dr Derek Ratcliffe and Dr Adam Watson. George was eager to bring in the birds in 1968; he had estimated the cost to the RSPB to be £750, but producing a film for television would be extra. He was cutting it fine but got final approval from the Nature Conservancy in June.

Fair Isle was chosen for the project because of its remoteness, the abundance of prey species and the fact that the experiment could be conducted by the resident observatory staff. It is a beautiful but isolated island, measuring about 5 kilometres from north to south and 3 kilometres wide, lying between Shetland to the north and Orkney to the south. It's a rugged island with big cliffs, but the more fertile southern part was where the 70 or so islanders lived on their crofts, cultivated their crops of oats, potatoes, turnips and hay and kept a few cows and hens, as well as sheep. Lighthouses at the north and south ends of the island were each occupied by three families when I lived there with my own family in the 1960s. The northern end of the island was heather moorland, rising to the top of Ward Hill at 232 metres, and this was where the islanders kept the majority of their sheep. We lived at the Bird Observatory, based in old naval huts which nestled in the slope running down to the harbour where the island boat, *Good Shepherd*, was moored. In summer the island was thronged with seabirds, with thousands of guillemots, razorbills, puffins and kittiwakes on the cliffs, and great and arctic skuas on the land.

Once we knew that the sea eagles were to arrive in late June, it was my job to work out how to look after them and then to release them. There was no handbook on what to do, so I drew on my knowledge of looking after other species as well as a bit of agricultural know-how and decided to house them in four large cages, each 3.6 metres square and 1.8 metres high.

Fair Isle Bird Observatory, the author's former home at the old naval station at North Haven; distinctively shaped Sheep Rock in the background.

Subsequently, we called them 'hacking cages', as the rearing in captivity of raptors for falconry and releasing them to fly was known as hacking. The cages were made of wooden frames and wire netting and they were built in pairs, each cage containing a small roosting shelter, a log of wood and a central perch. The first two were built by my assistant warden, Tony Mainwood, and myself on the hill slope of Erne's Brae, just south of Ward Hill, facing eastwards. At the observatory we were well used to this sort of work, as we often had to repair our big Heligoland traps, also made of timber and wire netting, which we used for catching migrant birds for ringing and research. We built the second pair of cages on the cliff top of Roskillee, just north of the observatory. The cages were made in such a way that, when the young were ready to be released, the front of each individual cage could be lowered down to allow the eagles to leave. Before building the cages, I had made the decision to keep the eaglets apart in case there was fighting between them, but a male and a female would live adjacent to each other in each set of cages. Soon we were ready and eager to start.

Johan Willgohs, the authority on the species in Norway, was ready to collect four young eagles once the Norwegian government had given permission for the birds to be collected and exported to Scotland. Sea eagles, along with other large raptors, had been habitually destroyed in Norway, much to the concern of conservationists, and Johan advocated that the species should be protected. He and his wife Einey visited eyries in Trondelag in northern Norway in the spring and returned in mid-June to collect four young eagles, two males and two females, from different eyries. The Loganair Islander aircraft, based in Orkney to run the new inter-island air service, was chartered for the occasion and, on 24 June, flew back from Bergen to Kirkwall airport with Johan, Einey and three of the young birds – the fourth had been delayed by a late arrival of a coastal ferryboat in Norway. The regular pilot, Captain Andy Alsop, soon set off for Fair Isle and landed the plane on the island's gravel airstrip. The eaglets had landed.

A few days earlier, George Waterston had come to Fair Isle for the momentous occasion, along with John Arnott from BBC Scotland and Dennis Coutts, a photographer from Lerwick who was a friend of ours and a keen birdwatcher. Also staying at the observatory, by chance, was a young friend of mine from

Inverness, John Love, who was later to become the main player in the sea eagle saga. We unloaded the first large cardboard box from the plane and opened the lid, to be confronted by the first of the massive eaglets. What awesome birds they were: huge, solid birds with big sharp bills and large feet and talons, and dark soulful eyes. Since then, our reaction at the sight of those birds – they are huge! – has been repeated again and again, every time a new person peers into a cardboard box to see their first young sea eagle. Once we had them unloaded, we took a pair in the

Arrival of the first Norwegian sea eagles at Fair Isle on Loganair Islander aircraft in 1968. From left to right, author, Johan Willgohs and George Waterston.
(*Photo by Dennis Coutts*)

observatory van to Erne's Brae, where each bird was put on an 'eyrie' or nest, built inside the shelter, with a supply of cut-up rabbit. The first two had been named Ingrid and Jesper by Johan and Einey; the third eaglet, a female called Torvaldine, was placed in the cage at Roskillee.

Over refreshments back at the observatory, Johan brought me up to date with each bird and its character, and gave me expert advice on how to feed them. With great excitement and anticipation, we returned to the cages that evening and fed each eaglet by hand with fish and pieces of rabbit. We immediately recognised that each had a different nature. Torvaldine was a noisy and bold female, who fed readily from our hand and was able to tear up large pieces of food; Ingrid was quieter and would pick up and eat pieces of fish placed in front of her; Jesper, the male, adopted a cowering posture and refused to have anything to do with us. That first day I force-fed him by opening his bill and pushing in some fish, but we decided to leave cut-up food in his cage and, next day, found that it had all been devoured. We were learning our first lessons.

Those first few days were busy times, with radio interviews for the BBC's *Afield* programme, recording for a *Sea Eagles for Fair Isle* film by the RSPB, photo sessions and a chance for island and observatory friends to view our new charges. The fourth eaglet arrived on a Scandinavian Airlines flight to Glasgow airport at 50 minutes past midnight on 6 July. George wrote to me: 'I collected it and it slept in my dormobile with me overnight in the car park, and after giving it a breakfast of fish I put it on the Shetland plane at 10 am. If it's a female call it Einey or Johan if a male.' It was looked after at Sumburgh and put on board the *Good Shepherd* on 9 July. It was a male, so the newly named Johan was put in a cage next door to Torvaldine. Our small, precious brood was complete.

I was eager to move to the next stage because I had decided the eaglets had two key requirements: plenty of food and as little human disturbance as possible. It was important that they remained wild, so human contact was to be minimised. As soon as they were able to tear up fish and dead rabbits, we started to cut down on our visits. We gave each eaglet about a kilo of meat or fish per day but we found that, usually, they didn't eat it all. In the first few weeks they spent a lot of their time lying down in the shelters or in the sun. They roosted in their shelters at night and were in them well before dusk; in fact, we often put two or three days' quota of food in each cage in darkness, so that the birds wouldn't see us and become used to our presence.

Each day, one of us would hunt rabbits or secure fish to feed the eaglets, and they grew rapidly. By 7 July, the first eaglet was seen perching freely on its log; on the 13th, Jesper was doing wing-flapping exercises, and he made his first flight across the cage seven days later. By the last week of July the three older birds were flying well, while Johan, the second male who arrived later, was younger and less advanced. They became shyer when they started to fly, and Torvaldine stopped calling at us when we approached. The main reason I had made the cages just 3.6 metres across was to lessen the risk of damage to the birds when they started to fly, because the further they could fly, the heavier their crash against the wire netting. With a short distance, they could not get up speed but could flap their wings to exercise. After two weeks, the restlessness slowed down and they spent lots of time preening their feathers and watching the world outside.

Once all the eaglets had been successfully reared to the flying stage, plans were made for their release. Aggressive skuas were nesting on the hill, so we delayed the releases until they had left for the winter. I did not want a mob of bonxies (great skuas) chasing one of our precious ernes over the cliffs and out to sea. I also suggested that one eagle should be released and allowed to settle down on the isle before another be set free – as usual, hedging our bets. Tony and I caught up each eagle for colour ringing and to check its weight, measurements and condition. While handling them, we were struck by the distinct warmth of their legs, feet and bills, which was quite unlike any of the many other species that I had handled for ringing.

The following table gives the weights, measurements and colour rings of the four young sea eagles:

Sea eagle stats, Fair Isle, 1968

| Eagle | 1 | 2 | 3 | 4 |
|---|---|---|---|---|
| Name & sex | Ingrid (F) | Jesper (M) | Torvaldine (F) | Johan (M) |
| Date | 14 Sep | 14 Sep | 30 Aug | 30 Aug |
| Colour ring | red | orange | green | blue |
| Ring leg | right | right | left | left |
| Wing | 690mm | 655mm | 725mm | 620mm |
| Wing on arrival | 440mm | 445mm | 410mm | – |
| Bill from cere | 57mm | 50mm | 59mm | 47mm |
| Bill depth | 38mm | 31.5mm | 39mm | 33mm |
| Tail | 380mm | 350mm | 380mm | 340mm |
| Weight | 5.8kg | 4.4kg | 5.5kg | 4.4kg |
| Weight on arrival | 5.2kg | 4.2kg | 5.5kg | – |

The difference between the large females and the smaller males was very obvious, especially in their weights and the size of their huge bills. When they arrived they had been nearly at their full body weight, so the main difference was in the big growth of the flight and tail feathers. I was delighted that their feathers showed remarkably few starvation marks, which are caused by a day with no food, apart from a few small ones which corresponded with the time when they were first collected from the eyries.

By the evening of 15 September we had fixed a release mechanism on the side of one cage with a pull-cord running across the Mire of Vatnagard to a hide, from which we could observe the eagles. From under cover, I could pull away one side of the cage from a long distance away, allowing Ingrid to come out of her cage of her own accord without being frightened. The 16th of September was a fine, clear day with no rain, so just after dawn – crouched in my hide – I pulled away the side of the cage. The others watched from the airstrip, much further away. Ingrid looked at the opening for five minutes, and then, rather than flying, out she walked. She stalked slowly up the hill, stopping for short rests to survey the scene and, 25 minutes later, reached the top of the slope about 100 metres from her cage. A raven flew in and landed beside her. Almost immediately the raven flew off and the eagle followed, the two of them soaring over the hill for several minutes before the eagle landed – rather clumsily – in the same place. Ten minutes later she was off again and soared over Skinner's Glig, only to be mobbed by a peregrine and chased towards North Felsigeo on the west cliffs. It was amazing for us to see this huge bird, after an absence of well over a century, soaring over Fair Isle. It was also a new experience for the birds of the island, which suddenly found themselves with such a large neighbour. Later that day, on my bird migration survey, I called by Upper Stonybreak to chat to my friend Georgie Stout, whose croft faced Ward Hill. 'I've seen your bird!' he said. 'Looked just like a barn door flying across the sky!' Little did he know that his description would go down in illustrative ornithological speak and is still used to this day.

And so began a period of intense observation, of learning how these birds behaved in the wild. Ingrid was regularly mobbed by ravens and hooded crows, both in flight and on the ground, while fulmars just glided beside her over the cliffs. She did not come back to the cage for food but searched the beaches instead, and was seen eating a freshly killed oystercatcher and great black-backed gulls shot by an islander. With a spell of good weather, Jesper – in the Erne's Brae cage – was released at 5.15 pm on 2 October; he flew out of the cage and was filmed by Dennis Coutts heading to the northwest cliffs. The next day he was flying with Ingrid. At 12.20 pm on 4 October we released Johan. He flew straight out of the cage and stayed in the air for 30 minutes, soaring over Ward Hill. On 5 October, all three eagles started to return to Roskillee to feed on carrion left in or beside the empty cages. The last of our sea eagles, Torvaldine, was set free at 2.40 pm on 20 October. She landed outside the cages and then flew off low over the sea towards the North Lighthouse. We then moved the food dump of carrion further north, to Wirvie.

These were exciting days for us and for the many birdwatchers who had arrived for the migration season. As well as seeing rare and exciting bird migrants, they were treated to the spectacle of these huge sea eagles soaring up over the cliffs. It was memorable to be part of this fantastic project.

In late October we had some strong northeast gales and the eagles hunkered down in the sheltered cliffs on the west side of Fair Isle. Although we did not see any of the eagles kill their own prey, Torvaldine was observed close to an injured greylag goose. Later, the dead goose had been carried away by an eagle. From this time onwards, they preferred to carry away food, to eat it in the cliffs.

By the end of the month we realised that Johan had left the island. Eddie Balfour, the Orkney RSPB man, reported to me that a sea eagle had been seen there but he had not confirmed the sighting. Throughout November and December, the three remaining sea eagles spent most of their time in the high north-west cliffs, so we placed food dumps at Erne's Brae and Toor o'da Ward Hill, the latter involving a hazardous scramble down the cliffs. It was a great boost to the project that my assistant, Tony Mainwood, agreed to overwinter at the Bird Observatory so that he could supply the birds with dead rabbits and gulls during the winter months, as well as monitor progress. He saw the birds collecting the food most days in December, and a pellet containing fish bones showed they were scavenging as well. On fine days they would soar and play together in the air, often giving long, shrill calls. These tremendous flying displays, especially over the cliff edges, were carried out even in the strongest winds. We were thrilled at how well they were fitting into a life on Fair Isle.

During January and February they were seen most days, with all three eagles seen on 33 days. The food dumps on the Tour o'da Ward Hill and Erne's Brae were maintained throughout the two months and the following carrion was provided: Ninety-three rabbits, four gulls, three lumps of mutton, most of a small grey seal carcass, a feral cat, five guillemots, five shags, one fulmar and one puffin. The rabbits were shot with a .22 rifle, while most of the remainder was found washed up on the beaches and transported to the hill. In February the three birds were seen together more often in the air and they developed a habit of flying down the west cliffs, across Meoness and up to Sheep Rock each morning, searching the beaches for carrion. In the third week of February they were seen regularly in the Sheep Rock area, in fact on several occasions perched on the highest point, no doubt exactly where the ernes perched in ancient times. In the last week of the month all three visited Hesswalls to feed on a large grey seal carcass washed up on the beach.

I returned to the island on 25 February after our winter break on the mainland and, as the plane landed us at the airstrip, saw all three eagles flying over the hill. I went out immediately with Tony to witness their fantastic aerial abilities. These are birds which can glide powerfully along the cliff tops on even the stormiest days. The main display consisted of the male diving down to a female in mid-air, the female turning on her back to present her outstretched talons to the diving bird. Occasionally they would touch talons and a few times were seen to spiral down, talons linked, for a short distance. Once I saw the male eagle do a full roll with fully spread wings. During these displays, the eagles – probably just the females – gave a long squealing call rather like a young razorbill calling from its nest. This call was also heard from the birds when they were perched together on the ground.

It was also fascinating to see how they interacted with other creatures on the island. Although the eagles came to the food dumps and carried away rabbit carcasses – some of which had been placed in the position adopted by crouching live rabbits – there was no evidence that they ever attacked or attempted to kill a live one. The island rabbits still showed no fear of the eagles soaring overhead and did not hide or bolt down their burrows. Neither did the eagles attack any sheep, although once, on Malcolm's Head, an eagle was seen to hover over a flock in an inquisitive manner, while Torvaldine, the least shy bird, would hover over any dog running about in the open on the hill. The eagles were no longer bothered by crows or ravens; in fact, they were dominant when all three species were at carrion, but peregrines would chase them when they flew near their nest site, south of Sheep Rock.

In early March, with a southeast wind, oiled seabirds, mainly guillemots, came ashore, dead or dying, which in those days was not unusual. On the first day of the month we found 25 guillemots on the beaches, and another 25 the next day. At this time, the eagles quartered the southeast-facing geos and carried up seabird carcasses from the tideline. The female eagle, Ingrid, left the threesome after 6 March and led a rather solitary existence in the west cliffs, where she found all her own food. Torvaldine and Jesper went regularly to the Erne's Brae carrion dump and Jesper often fed on the female's carrion after she was finished. In

Female sea eagle, Torvaldine, carrying off dead rabbit from carrion dump at Ward Hill, March 1969.

the early part of the month I built a hide at Erne's Brae and from it photographed the sea eagles and ravens. By this time, I thought they were finding half their own food in the form of carrion. On 27 March one eagle was seen hovering over a dead kittiwake in the sea off Hesti Geo but did not pick it up.

On 12 and 20 March, Torvaldine was seen hovering over sheep at Setter and Erne's Brae but made no attempt to attack them or land close to them. On 12 March she was perched on the cliff opposite Erne's Brae while I was in the hide. I saw one of the islanders driving his tractor across the airstrip, nearly half a mile away, his collie dog running alongside. I noticed Torvaldine bobbing her head and looking inquisitively in that direction; almost immediately she took off, flew right over to the airstrip and hovered over the dog for a few minutes before returning to her perch on the cliff. The behaviour seemed to be purely inquisitive; she was not hungry, as there was plenty of food available at the dump. On one occasion, two eagles were seen washing in a freshwater pool on Sukka Mire.

On 23 March 1969 I wrote to George asking that we replace our lost male with a new one from Norway in the summer. In a two-page letter, I also said: 'if the mystery bird in Orkney proves to be Johan we should trap it and take it back to the isle. If we lost our second male, it would be best to trap a sub-adult or adult in Norway. I'll write and sound out Johan Willgohs. I also think we should give the experiment a broader base by establishing a second reintroduction site away from Fair Isle. I think Hermaness in Unst, Fetlar or Fitful Head and Sumburgh – all in Shetland – are my favourites. If we could get permission I'd recommend four new eaglets this summer for Shetland and a replacement to Fair Isle. Last year we were offered some captive birds from the continent but after my observations of our sea eagles I am very much against using any sea eagles which have been caged for any length of time. I have noticed a difference in behaviour between Torvaldine, who was caged for the longest time, and subjected more to visitors, and the two eagles at Erne's Brae, which are much shyer. In future I would strongly advise that eagles in cages be left strictly undisturbed and visitors are asked never to go within view of the cages, and that feeding visits are kept down to once every three days or so.' All my suggestions were based on one main principle: keeping the birds wild.

For the whole of April, Torvaldine and Jesper were together, usually in the northwest cliffs, where they roosted. Ingrid was last seen on 12 April at Lerness and Guidicum; she appeared to be in excellent condition and I was certain she had decided to leave the island. Sometimes, on good days, the eagles would soar very high over the island, with Orkney and Shetland clearly visible about 40 kilometres away, and I wondered if they might even, at the clearest times, have seen the clouds over Norway, 375 kilometres to the east. In late April I received two unconfirmed reports of a large eagle north of Whalsay and at Vidlin in Shetland, but it was not properly identified.

Dead rabbits and seabirds were provided at Erne's Brae throughout April and May but the sea eagles came less regularly; maybe the skuas, which had returned to breed, were too aggressive for them. Of course we wanted to know what would happen when lambing of the small Shetland sheep started, so we kept a careful watch on the hill, where the crofters were frequently out working with their sheep. It was encouraging to record that there was not a single report of the eagles killing a lamb during the lambing season on the island. In fact, there was not even one observation of them eating or carrying away a dead lamb.

On 8 May we gathered the first evidence of an important step in our experiment when a Mr and Mrs Lienart (two Belgian birdwatchers staying at the Bird Observatory) watched the two eagles soaring over the west cliffs at Skinner's Glig and saw one of them catch a fulmar in flight before releasing it, apparently unharmed. A little while later this episode was repeated; the eagle carried the struggling fulmar for a few seconds and then let it go. On 10 May I was at Erne's Brae and disturbed both eagles from Matchi Stack; as they flew across the geo to Lerness, there was tremendous panic among the fulmars and kittiwakes, and to a lesser extent the puffins. Eight days later, I saw both eagles half-heartedly chasing fulmars at Guidicum, but on the early morning of the 20th I found the remains of two freshly killed fulmars at one of the eagles' favourite perches on the high cliffs above South Felsigeo. Only the bill, legs and feathers were left, and it was obvious that the young eagles had mastered the technique of catching adult fulmars in flight, killing and eating them.

Very early on the morning of 20 May, a beautiful day, I was walking from Wester Lother towards Dronga, in the far north-west of the island, when I saw both eagles soaring over the cliffs; at times they were mobbed by a hooded crow. With my binoculars I could see that Torvaldine was carrying something in her talons and trying to peck at it with her bill. About 15 minutes later, when I reached the high cliffs above Felsigeo, I could see that she was carrying a clear glass bottle, about a half-pint size, by the neck. As I watched, she flew high over Felsigeo still trying to peck at the bottle, but lost her grip, the bottle dropping 250 metres into the sea. I thought she had probably seen it glistening in the water and had picked it out, perhaps thinking it was a silvery fish. Torvaldine's moult was now advanced and only six old feathers remained in her tail.

That afternoon, a sea eagle twice flew past the Bird Observatory; it was repeatedly mobbed by gulls and crows over

Buness. Later it was at the Sheep Rock and then flew across the hill, chased by skuas, to land in an exhausted state on the north shoulder of Burrashield. We could not get close enough to identify the bird from the colour ring combination, but it was a female and, because of its exhausted condition, we thought it was Ingrid. The weather that day was very good, with excellent visibility to Shetland and Orkney. On the following days, only the usual pair was present, though, so maybe she had returned to the isle and then left again.

June started off very well, with both eagles seen on the 1st and a freshly killed fulmar found near Erne's Brae. When disturbed, the two eagles were mobbed by herring gulls off the Tour o'da Ward Hill and then flew south to Troila Geo. The pair was seen in the northwest cliffs each day, up to and including the 7th. We then had a spell of very good weather, with excellent visibility, and we realised that Torvaldine had gone. Although none of the three sea eagles was seen leaving the isle, it is worth mentioning that on occasions the eagles soared very high – up to 600 metres – over the island and out over the sea, for up to 5 kilometres, before returning to the cliffs. So now we were down to just one young sea eagle.

Throughout June, Jesper was observed in the very high north and west cliffs; he was by this time in heavy moult, which was normal. On 29 June, Jesper was seen at Easter Lother eating a fulmar in the cliff; on 2 July, an eagle flew over the shoulder of Ward Hill, while on 3 July he was disturbed from the seabird colony in a rock pile on the beach at Wester Lother and flew off towards Millens Houllan. A search showed that two young shags had been eaten on their nests.

Then followed nearly six weeks without a sighting, but Jesper had obviously led a secretive life in the north cliffs because, on 13 August, he was disturbed from the large sea cave at Lericum, just west of the North Lighthouse. He was seen in the same place on 16 and 18 August, and fresh kills of young fulmars, shags and gulls were found on the beach. On 19 August, Gordon Barnes, a crofter on Fair Isle, climbed down to Lericum beach and the sea eagle flew out of the cave, tried to land on an offshore skerry but fell into the sea. It flapped and struggled in the water but could not become airborne; instead, it drifted ashore where it was caught and examined. It was Jesper, fat and well fed, but with his whole plumage soiled, each feather matted and smelling strongly of fulmar oil. The fulmar has a defence mechanism of spitting really revolting oil at any approaching predator – or bird ringer – and over the years we had found raptors such as peregrine falcons and honey buzzards completely incapacitated by it. This appeared to have happened to Jesper: each time the eagle had approached a young fulmar on a nest ledge, he would have been spat at by the bird, before and during the struggle to take it as prey. The eagle was carried up the beach and left on the rocks. He was seen in the same geo – narrow inlet – on 20, 21 and 22 August, but by 28 August he had moved round to Easter Lother, where there was a dead gull on the beach. This was the last sighting of a sea eagle on the island, and I concluded that fulmar oil had led to his death, probably through drowning in the sea.

It was a sad end to our project. On the positive side, we had demonstrated that young sea eagles could be safely translocated from Norway to Scotland and successfully reared in cages without becoming tame or damaged. We had designed an effective hacking cage and found that the birds learned quickly to tear up their own food in captivity. We had shown that the eagles could be successfully released and attracted to food dumps over long periods, to encourage them to overwinter at the release site. They had learned where to roost and the best beaches for

A fulmar defends its egg by spitting pungent stomach oil at an intruder.
(*Photo by Annelies Leeuw/Alamy Stock Photo*)

finding carrion, and they had frequented the exact places where the species nested long ago. They had learned to kill seabirds, but obviously would need to learn how to deal more effectively with fulmars, as there was no doubt that oiling from the young could cause serious problems.

At the time I was optimistic that some of our sea eagles might be alive in nearby islands, on mainland Scotland or even back in Norway, and we now know that young Norwegian sea eagles released in Scotland can return to their homeland to breed. I received unsubstantiated reports from northern Scotland and the Outer Hebrides, and a strong lead led me to make a visit to Whiten Head, in north Sutherland, in late March 1969, but my day's search revealed no sea eagle.

I was convinced that it would be well worth continuing the experiment by bringing in more birds, but I was no longer sure that Fair Isle was the best choice. It had many advantages but it was small, and suffered from a lack of secluded cliffs where the young eagles could peacefully roost, eat their food and moult without being continually disturbed and mobbed by gulls and other species, such as great skuas. It was some consolation that the law in Norway had been changed and that sea eagles were now protected; it is satisfying to think that the experimental reintroduction at Fair Isle helped to influence this decision. I was, though, sad that I had failed to help realise George's dream of having sea eagles breed once more on Fair Isle. His efforts to get them there were Herculean and, sadly, he died before knowing that his project paved the way to final success.

In 1969 I was engaged in a massive project to build a brand-new bird observatory at Fair Isle so we could move out of the old naval huts, which had been in use since 1948. The following year was a very busy one, as we brought the new observatory into commission and welcomed hundreds of visitors. That was my last year there; in the winter, I moved with my family to the mainland and started a new job with the RSPB in the Highlands.

## Successful Nature Conservancy reintroduction, 1975–85

While sea eagles necessarily became less of a priority for me, I had now, like George and Pat before me, been bitten by the bug. In my view it wasn't a matter of 'if' sea eagles would breed again in Scotland – it was 'when'. In the RSPB there was no appetite for reintroductions; the subject was, in fact, off limits, as it was with much of the ornithological establishment in Britain. Reintroductions were simply taboo. But this didn't stop those of us who still wished to realise the dream from talking it over when we met on field trips or at meetings. The Nature Conservancy staff were really kind to me when I started working for the RSPB in the Highlands in 1970. I was a lone worker and they took me under their wing. The Aviemore staff helped me with office space and, when I moved to the Black Isle (not an island at all, but an area across the firth north of Inverness) a few years later, I became firm friends with Andrew Currie, their man in Easter Ross, as well as with the staff at the headquarters in Inverness. They included Niall Campbell, Martin Ball, Peter Tilbrook and Sandy Maclennan, and we often worked together on conservation matters, especially relating to the arrival of North Sea oil, a massive challenge to nature.

Sometime in 1974 it was whispered to me that Dr Ian Newton at the Nature Conservancy headquarters in Edinburgh, with support from the chief scientist Dr Derek Ratcliffe, was working on a new sea eagle reintroduction proposal. My diary tells me that on 18 October I talked with Martin Ball and Niall Campbell in Inverness, because they wished to know all about my experience with sea eagles at Fair Isle, while on 11 November 1974 I visited the Nature Conservancy head office at Hope Terrace in Edinburgh to talk with the director John Morton Boyd, Tony Colling (the species licensing officer) and Ian Newton about their sea eagle proposal, as well as about goshawks. Morton was an old friend of mine. He had taken me on grey seal expeditions to the Hebrides and North Rona in the autumns of 1961 and 1962 and we had become great friends (despite him pulling rank on me in a game of Scrabble when camping on North Rona over my game-clinching use of the word 'zany'). My problem, though, was that my organisation, the RSPB, was not in favour of reintroductions, so I had to keep a low profile and make certain that my contribution to any discussions was a private one.

The decision to move forward was made and Martin Ball was tasked with making it happen on the island of Rum, the Nature Conservancy National Nature Reserve in the Inner Hebrides. Martin was a near neighbour of mine on the Black Isle, so we regularly chatted, often on the Kessock car ferry to Inverness, like something out of a John le Carré novel. One day, I mentioned to him that John Love was back home in Inverness from Aberdeen University and looking for an interesting challenge. I was delighted when John got the job of sea eagle project officer. Suddenly, it was all go. On the evening of 2 May, Martin and Sheila Ball and John Love came round to my home to talk about the Fair Isle project. The Nature Conservancy went on to build cages on Rum similar to my Fair Isle ones, Martin went on a recce to Norway with Johan Willgohs, the RAF had agreed to transport the birds from Norway, and John Love went to Bodø in June to help Johan collect them. John Love has written a full history of the project in two superb books, *The Return of the Sea Eagle* and *A Saga of Sea Eagles*.

I was very envious and wished I had been in Norway with John, but on 26 June 1975 Martin took me with him to RAF Kinloss, not far from where I lived, to collect the young sea eagles. In the 1970s and '80s, Kinloss was a very busy Air Force base, with the big maritime surveillance Nimrod aircraft regularly patrolling the seas off northern Norway as part of the NATO Cold War operations. RAF Kinloss also had an active ornithological society (RAFKOS) with some really good birdwatchers. I was often asked to give talks at their monthly evening meetings and became good friends with many of them.

We were taken into the base at 5 pm and met the commanding officer and his staff; the plane was due at six, so we assembled near the apron to watch the big grey plane come in from the sea under an overcast sky and thunder down the runway. Once it was parked, four big cardboard boxes were carried down the steps and we were able to open them and inspect the eagles. They were bigger than the ones I had received on Fair Isle and all were in excellent condition. We loaded them into Martin's vehicle and drove them to his office in Inverness, where we fed all of them on fish. In my view they were four very fit young sea eagles. Martin and I then set off to Rum and I learned that John had to find a much slower way home, by commercial aircraft, as only the birds were allowed to fly with the military in that first year. We had a midnight sailing from Mallaig, with lots of sightings of Manx shearwaters low over the sea, arriving at Rum at 1 am. Peter Corkhill, the nature reserve manager, was there in the dim light of the Highland summer to transport us to the eagle cages. The young eagles were placed in their new quarters, given fresh fish, and we went back for a hurried sleep before catching the ferry at 9 am. As we drove back to my house on the Black Isle, I felt absolutely thrilled that a new and bigger project had started, and that this time there would be success. By means of phone calls and chats, I gave positive support whenever asked.

The next year, 1976, John and Harald Misund collected 10 young; and after meeting John at RAF Kinloss I took some of these eaglets in my car while the others travelled with John and Martin. I set out to Rum in the evening, and after putting the birds in the release cages I stayed overnight with John at his house overlooking Village Bay and checked the eaglets first thing. It was great to see 10 healthy young, a really meaningful step in ensuring success in the future. That year I returned on 20 September with Ian Newton and Len and Pat Campbell, university friends of John's; we had an interesting boat journey to the island via Eigg and Canna, landing at Rum at 4 pm. We held a sea eagle group meeting in the amazing Kinloch Castle in the evening, and next morning we viewed the eaglets. My notebook records that we reckoned there were three males and four females but that we were unsure about the others. (John later proved that there was an equal number of each sex.) Even at this stage, though, we were always looking to the future and at how to broaden the project out in the years to come. We had discussions about holding a small number of eagles for captive breeding; on 30 January 1977 John gave an excellent lecture at the Scottish Ornithologists' Club annual conference; and the following afternoon we all assembled at the Nature Conservancy HQ in Edinburgh to plan the next steps.

Ian Newton and author viewing young sea eagle on its shelter at Rum, 1976. (*Photo by John Love*)

The next year, on 6 March 1978, I was at a meeting at the Aviemore office with the Nature Conservancy team and Doug Weir, of the Highland Wildlife Park, discussing further the ideas of captive breeding on Rum, or with a couple of captive birds at the Wildlife Park. This plan was later abandoned as it would have been very difficult and was not endorsed by the Norwegians. On 20 June I was invited to join the collecting trip to Bodø in North Norway. It was a 4 am start, so that I could attend crew briefing at RAF Kinloss and depart in the Nimrod reconnaissance aircraft at 8 am. These flights were operational sorties, with the collection of sea eagles a kind of add-on, so the crew spent several hours searching for submarines and boats in the North Sea before heading north. We could move about the plane and frequently ended up in the galley for a chat and tea. When we landed at the Norwegian Air Force base at Bodø, I met Harald Misund for the first time, and Morton, who had arrived earlier. Harald, who was a Norwegian Air Force officer and the local sea eagle expert, took us into the hills and along the coast where we viewed four different white-tailed eagle breeding locations – the wild flowers were plentiful and beautiful, and in the pale sunlight, close to midnight, redwings and fieldfares were singing. It was really good to meet Harald, as without him the project would not be progressing so well. As a serving officer of the Norwegian Air Force, he made arrangements at Bodø air base easier, but I did note his wish that we should let all the eagles fly free as soon as they were ready: he did not like them held in captivity. We had a late night and an early start when the eight eaglets were loaded onto the Nimrod. It was a high-level flight home and from 9,000 metres I could see the oil rigs in the Shetland Basin and then my old home, Fair Isle – where we had first reintroduced the sea eagle – sunlit in the ocean. Martin and my RSPB colleague Roger Broad then took the eagles to Rum.

By 1978 the first of the released sea eagles from Rum were starting to explore Scotland, with two on the Isle of Canna and one even roaming as far as Loch Ruthven near Inverness. In cooperation with John on Rum, at my RSPB office near Inverness, we started to receive sightings and we encouraged people to look out for the birds, with appeals on local radio and in the papers. The famous Caithness Glass company in Wick engraved a sea eagle paper weight for the project to gift to Harald when John went to receive six eaglets in 1979. Another eight young were reared and released on Rum in 1980, another five in 1981 and 10 again in 1982, when the Norwegian Air Force brought them over to Kinloss, as the RAF were fully engaged with the Falklands war. In 1981, the Nature Conservancy, in view of funding difficulties and the need for outside funding and wider representation, set up a more formal white-tailed eagle steering group. That committee is still active, but of the original members only John and I remain.

The year 1983 was a special one for me with the sea eagles. We were sure we might be missing the first attempts at breeding, so the Nature Conservancy hired a sailing boat from Strollamas on Skye to carry out field surveys. On 14 June we sailed the yacht through the Hebrides for a week, cooking on board and sleeping in sheltered bays. It was a great expedition, with John Morton Boyd, Peter Tilbrook, Martin Ball, acting as skipper, Julian Hunter, John and myself. We searched the cliffs and islands from South Rona to Lord Macdonald's Table north of Skye, and on to the Ascrib islands. The next day we sailed across the Minch to the Shiant Islands, where sea eagles used to breed. The seabird colonies were very impressive, especially the amazing razorbill nesting grounds on the long boulder beach. We continued to the Isle of Harris, where we anchored for the night in a big sea loch. On the 17th we had an exhilarating sail back across the Minch in rough seas, to make landfall for the night at Dunvegan. We sailed south along all the big cliffs of Skye; Julian, who had a yachtmaster's ticket, said it was the most exciting – and by that he meant frightening – sailing trip of his life, as I was always wanting to get closer to the cliffs (and the rocks) to search for eagle eyries. In this way, we saw several pairs of golden eagles and their nests. After a night at Carbost, we sailed for Coruisk, seeing several more cliff eyries of golden eagles as well as a very rare white-billed diver on the sea. The massive north cliffs of Canna dwarfed us as we watched a pair of three-year-old sea eagles displaying above us. It was a beautiful evening, so we circumnavigated Rum, which paid off, as we found a pair of adult sea eagles eating a dead gull near a starter nest site at Sgurr na l'Iolaire, or Eagle's Peak. We spent our last night at Rum and, next day, in deteriorating weather, we returned the boat to Skye, for our colleagues to make a second week's search further south. Fieldwork can be exhausting, and sometimes doesn't prove much, but it's nearly always exhilarating. And it's the only way, truly, to understand nature.

A month later John invited me to join the collecting visit to Bodø and we flew out with the RAF on 13 June. We did not land until 5 pm, as the crew made a very long search for Russian submarines over the north Norway seas. John and I dozed in the back, waking only to see Mick Muttitt walk by with an extinguisher to check a small electrical fire. We landed safely, though, and Harald Misund collected us for an evening walk in the local woods. The next morning we were away by 8 am to visit areas to the north; the woods were ringing with birdsong and we saw four nest sites of sea eagles, although the only easily accessible nest was empty. At 4 pm the warden took us by small boat to the Kjerringoy islands, where we collected one young from a brood of two in a very small birch tree. Two adults were tandem flying over a nest site round the next headland, but there was only one young, which Harald ringed but left in the nest: we could only take a chick if there were at least two in the brood. As we were crossing a sound to search more islands, a flock of 15 graceful long-tailed skuas flew over, heading for their tundra nesting grounds. We visited several more eyries and ringed two young in a ground nest. As we walked across one small island a flock of twelve sub-adult and juvenile sea eagles flew out from their roost in a birch wood – it was a magical moment.

Finally, in the midnight sun, we got back to Harald's home after a fantastic, never-to-be-forgotten day – my sea eagle tally for the day was 46. The next morning, the last young sea eagle

John Love and author at sea eagle nest on ground on small island in north Norway. One young collected for Rum. (*Photo by John Love*)

was collected from a ground nest of two young on a small islet; one of the young had wandered three metres from the nest. Later we viewed the tidal maelstrom at Saltstruman Bridge, a favourite location for sea eagles in winter when they catch big fish brought to the water surface by the tidal upwellings. In the evening Harald showed us his summer cabin hidden in the mountains. It was a wonderful hike with lady's slipper orchids and globeflowers among a beautiful array of wild flowers, while bramblings and redwings sang in the birch trees. At Harald's beautifully built cabin we had tea from the most delicate china tea set, and then explored the mountains, where golden eagles and gyr falcons bred and moose wandered. It was idyllic, and probably the reason we did not get back to Bodø until after midnight. The next day we flew home to Scotland; John taking the eagles onwards to Rum.

Nineteen-eighty-five was the last year for collecting young in Norway, and Colin Crooke, my new RSPB assistant, went with John to Bodø. Harald had collected 10 young and they arrived at RAF Kinloss mid-afternoon on 20 June. They were all in great condition and brought the total imported to 85 young in 10 years. The Eagle Star Insurance Company had started to fund the project, and on 4 July I scrambled onto a big S-76 helicopter at Inverness airport, with my daughter Rona and her friend Nicky Crockford scrounging a free lift to the Uists. We refuelled at Stornoway and then met the Eagle Star private jet at Benbecula airport. On it were Sir Edward Mountain and Nicky Mountain of Eagle Star, as well as Council members of the RSPB with the Director, Ian Prestt. A happy surprise for me was that they had brought with them the famous French raptor expert, Michel Terrasse, whom I had last met at his vulture reintroduction project in the Cévennes. We had a quick flight to Rum, where John showed the group the new eagles ready for that year's release. After an impromptu picnic of smoked salmon on the castle lawn, the helicopter took us to Mull via Ardnamurchan. Roger Broad was there to meet us and we were taken to the watchers' tent in Glen Forsa, where Dave Sexton and Mike Madders showed the group the first Scottish-bred sea eagle chick on its eyrie, with the male close by. We flew back across the Minch to Benbecula for the group to return to London, while I was finally landed back in Inverness. It was one of the very special days in my life.

The collection, rearing and releasing of 82 young sea eagles in 10 years was a major undertaking, which involved huge amounts of work and dedication. Much credit goes to John Morton Boyd and Martin Ball of the Nature Conservancy, who through the government agency kept the project going for all that time. Morton must have had a difficult task in persuading the government to continue the funding as the years went by. The RSPB were heavily involved through my and Roger Broad's fieldwork, along with that of summer assistants – especially as the eagles started to settle into potential breeding areas – and enjoyed funding from the Eagle Star insurance company. Harald Misund in Norway was a tower of strength with his annual collection of young eagles; it was a major task each year, liaising with landowners and boatmen, monitoring nests, collecting and caring for the young while in Norway, and dealing with the worry of wondering if he could find enough young. RAF Kinloss and its Nimrod crews, especially 120 Squadron, were the other reason the project was successful; without them, getting the young eagles to Scotland would have been very difficult. The biggest credit goes to John Love, who for 10 years ran the project, organised the supplies of eagle food, cut it up and fed the young – a hard daily task in all weathers. John maintained the records and kept in touch with many contacts. He decided, each summer, which young were ready for release and then worried about what would happen to them. He successfully carried out a mammoth operation, and to him, and the others, we are indebted. Thanks to them, the sea eagle is back home.

## Claiming back their ancient haunts

While John was on Rum, running the project, the eagles started to explore Scotland. The first time I really felt that success was in sight was on 23 March 1981, when I was on my way to give an evening lecture at Plockton Hall in Wester Ross. I had taken the northern route from home near Inverness, and as I approached Shieldaig, I saw two hooded crows and a buzzard mobbing something on the rocky beach. I stopped my car and, with binoculars, was amazed to see that the alarm was being caused by a young sea eagle. I had time to see a green ring on the left leg before the bird took off and flew out to circle over the sheltered bay, Ob Mheallaidh, a southern extension of Loch Torridon. Suddenly I realised it was hunting, for below it in the sea was a mother otter with two big cubs. She was eating a fish, but each time the eagle swooped down to try to steal the fish, the otters dived. Long ago, I'm sure, ancient sea eagles would have stolen fish from otters in that very place. This young sea eagle knew what to do but it needed to learn, by experience, to be better at robbing fish from otters. Standing there, as dusk approached, with the great cliffs of Ben Sheildaig behind me and looking out over the dramatic Wester Ross scenery of sea lochs, islands and rocky headlands, I knew that the sea eagles were definitely back and that, sometime soon, they would breed.

I also knew, though, that there would be setbacks. A few weeks later, Hugh Clark, a schoolmaster in Wick, telephoned me with news of a poisoned sea eagle in Caithness. He collected it next day and it reached our office on 15 April. My assistant Roger Broad went off to search the location for illegal activities and I did interviews for Grampian TV and the BBC. The bird was a male named Erin, released by John the previous August, and it had eaten a dead hare laced with a lethal poison called Phosdrin. It was the start of too many losses to illegal poisoning carried out by shepherds and gamekeepers, where in many cases the indiscriminate and illegal poison was put down in baits to kill foxes, crows and large gulls, rather than eagles. On 18 November 1981, Pete Ewins and I searched an area near the MacLeod's Tables on Skye where a hill walker had reported a dead sea eagle. We found the long-dead remains of Z12855, which had been released on Rum on 18 October 1977. The death of a four-year-old was very disappointing. Nearby were the remains of a dead sheep and, further away, we found a dead herring gull by a stone dyke. We sent the remains away for analysis and visited the police station at Dunvegan as it looked to me like illegal poisoning. As with quite a few later cases, the evidence for a prosecution was very difficult to obtain, but publicity was one way we chose to try to reduce the risk to eagles.

In 1981, with funding from Eagle Star, the RSPB became a full partner of the project, carrying out surveys away from Rum. Colin Crooke joined my team in the Highlands and went to the Isle of Skye to check out sightings and potential nest sites, while Roger Broad did similar work on the island of Mull, where he had Jeremy Moore as the roving warden. Colin found his first sea eagle on Skye on 2 February, while carrying out a thorough search for golden eagle eyries in preparation for our first complete national survey of that bird in 1982. These two island groups would feature strongly in the sea eagle story.

Meanwhile, on 4 August 1981, Malcolm Harvey and I were in Glen Affric searching for honey buzzards when an adult sea eagle chased an adult golden eagle across the ridge towards Loch Beinnevean. It was probably trying to steal prey but soon came back along the ridge, being attacked by buzzards and a peregrine. We had not expected them to settle in inland glens, but on a golden eagle survey in the same area on 22 March 1982 I saw the same adult soaring 900 metres above Loch Affric at midday. Over the next half an hour I saw it several times, once being chased by two adult golden eagles. On 10 April I was there again, surveying golden eagles at the west end of Loch Affric, and on the way back had superb views of the sea eagle as it did a spectacular dive down to land on a big ledge on a rocky buttress above me. An adult golden eagle watched from above, but the sea eagle soon soared out and gained height, gliding over the whole glen, finally reaching about 2,500 metres in a blue sky, the white tail showing clearly. Sadly its aerial displays failed to find it a mate for its putative nesting site and, after a few years, we lost contact. Before it departed I was pleased to take John Morton Boyd to view the cliff and watch the sea eagle soaring out over Loch Affric. For him, one of the earliest believers in the plan to restore sea eagles to Scotland, it was a truly thrilling spectacle. As we stood on the trackside in the summer heat, two old friends, we felt great joy that the plan was finally becoming reality.

Roger Broad and Colin Crooke soon reported that things were happening on both Mull and Skye and that we needed to spend more time trying to work out which sea eagles were present. It was Mull that made the running, and on 1 December 1982 Colin and I left home early and caught the 9 am ferry to Mull. We met Roger and Richard Coomber on the Mull side and talked about the eagles, spending the day in the Loch Don area, first of all climbing the mountain side to the favourite crags of the sea eagles. The ledges were very wet but we found two eagle pellets and a dead hare. An adult flew over and looked at us before returning west and was lost in the mist. We checked a favoured roost, where there was a trial nest and more pellets; an adult flew out of the oaks, and later another full adult flew over Loch Don and did a short display flight. It was an eventful day with much to ponder on our long drive home.

In the winter we decided that the action was going to be on Mull, so Roger was assisted in 1983 by Roland Ascroft and David Bromwich. In February, at least five sea eagles were on Mull, with other singles in Knoydart, Kintail and Kyle of Lochalsh. As soon as they thought the pair was incubating eggs at the cliff site, I drove over and caught the ferry to the island, on 7 April. David collected me and we met the others at the viewpoint from where we could see an adult on the eyrie. We climbed the side of the corrie and watched the incubating eagle. I crossed the corrie and up into the snow at the far side to get a better view from 150 metres; the pair was suddenly above me, calling, so I retreated, the female returning to the nest and 'rocking' gently back down, showing that they definitely had eggs. The male was smaller and paler-headed, and while perched in the cliff engaged in much calling and head turning. He then flew 50 metres to a knoll – and imagine our amazement when he was joined by a second female, which billed with him. We scrambled back down the hill and watched another changeover at the nest. We were so excited that the first clutch of eggs had been laid, but anxious about what would happen with a breeding trio.

The wardens on Mull continued their watches and Martin Ball and I returned a week later. Roger met us off the ferry and we slogged up the hill to the watch point, where we found Dave watching the second female on the nest. She was sitting too high to be properly incubating, and the other female was perched on a rock. We watched for 15 minutes and then decided the nest needed to be checked, so we walked over and Roger scrambled onto the ledge at 1.45 pm. Neither female called at us, which was a bad sign, and Roger soon shouted down that one or two eggs were broken outside the nest, with two eggs still intact. The egg breakage was almost certainly due to the trio's antics while incubating. The nest – with the remains of about 15 mountain hares – was just a scrape in the earthy ledge and was very wet; in fact, it wasn't a properly built eyrie at all. Finally, after three days of rain and low cloud in mid-May, the nest was abandoned. What had been an encouraging start had ended in

Sea eagle nest in oak wood on the island of Mull, 1983. This would become the first successful breeding site in 1985.

failure, but that's where perseverance comes in, again and again. On 29 April another nest was found in an oak tree, where a younger female brooded a single egg, but that, too, broke during incubation.

Nineteen eighty-four looked more hopeful but Dave Sexton and Mike Madders had to weather yet more disappointments as the trio did not attempt to breed and the younger pair's single egg failed to hatch. I visited the nest site on 9 May with John and Roger to see the potential for making a film, but decided it was impossible because the oak tree was in full leaf. We checked the nest with our mirror pole and, disappointingly, there was just one egg. On 10 May John and I went to the Isle of Harris, where Andrew Miller Mundy, with Nigel Buxton, took us by fast rib round the remote rocky coast to a new nest. Alas, it was another failure: the pair were perched by the now empty nest. On 6 June I was back on Mull to show Willie Wilkinson, the Chairman of the Nature Conservancy, and a keen birder, the trio of adults displaying over Glen Forsa. There was also a pair on Rum, four birds on Canna, the nest with failed eggs in Harris and four younger birds in Shetland, where a young pair were showing interest in an ancient sea eagle site on Fitful Head. On Skye, Kate Nellist, a warden at the Youth Hostel in Glenbrittle, reported one non-breeding young pair with a nest in an unsuitable tree, as well as three other singles. On 20 August, her partner Ken Crane and I built a big new nest in a good tree, helped by local Forestry Commission staff. Kate and Ken were to become important eagle fieldworkers

I wished, of course, that it had been a successful year, but one finally arrived in 1985, when Roger, Dave Sexton, Mike Madders and Keith Morton saw two young in the eyrie. I was delighted to receive Roger's excited phone call. On 6 June Johan Willgohs arrived with his wife to stay at my home on the Black Isle. John and Brenda Love and Martin and Sheila Ball joined us for a happy evening of talk about sea eagles and Norway, and next day we drove to Lochaline and caught the ferry to Fishnish. Roger collected us, and with Dave and Mike we drove up Glen Forsa and then walked past the Highland cattle to the watchers' tent. The occupied nest was in an oak tree, and with the telescope we could see the adult and its single surviving young on the nest. We sat there watching and yarning for a couple of hours, Johan remembering the visit to Fair Isle in 1968 and the first years with Martin and John in Norway. The adult male flew out and soared above the nest tree just before we left for home. Johan remarked how very pleased he was that we had the great birds back breeding in Scotland, and for us it was a poignant visit which renewed our friendships with our Norwegian colleagues. In 1986 the first wildlife holiday companies made reference in their advertisements to sea eagle watching opportunities, and we had no idea then how big a business that would become. That year, our RSPB office near Inverness produced our first ever leaflet in Gaelic, featuring sea eagles, appropriately enough translated for me by a man from Mull.

When Colin Crooke went to carry out eagle surveys on the Isle of Skye in 1981, he soon met Ken Crane and Kate Nellist, the wardens at the Youth Hostel in Glenbrittle. Keen hillwalkers from London, they were soon out eagling in the field with Colin and in no time were our eyes and ears for eagles in Skye. We had great days talking about eagles when we met and, with a copy of my annotated golden eagle maps, they started to check out all the confidential eagle nesting sites. Since then, they have gone on to become two of the most experienced golden eagle field observers in the world.

Colin, Roger and I often talked about how we would know the project was going to be a long-term success. I came up with a benchmark: when the number of young reared in the wild in Scotland in a year matched the number that had been reared and released in a year by John Love. That number was 10. In 1985 the first young was reared; there were two in 1986, three in 1987, back to two in 1988. After a peak of five young from nine pairs in 1989, success fell back to only two young in 1990. We could not reach the target of 10 young, so we recommended that a second group of young sea eagles should be imported from Norway. I proposed this at the November 1991 annual meeting, to start the following year, with some to be released in Wester Ross and others somewhere between Easter Ross and Caithness. Rhys Green, a colleague in the RSPB, was asked to examine this proposal, and his scientific modelling and analysis showed that it was, indeed, advisable to release more young.

By this time I had left the RSPB to become a wildlife consultant, but retained a place on the Sea Eagle Working Group as an independent. I had also become a board member of the newly created Scottish Natural Heritage (SNH), following the break-up of the UK-wide Nature Conservancy Council. The regular meetings of the Working Group discussed the second-phase releases and, following scientific analysis of the Scottish sea eagle breeding performance and future viability, a decision was made to import a further 60 young. On 18 August 1992 we had an excellent meeting in Inverness to agree all the logistics for the new phase of the reintroduction. Before he caught his evening flight south from Inverness airport, I took Ian Newton to the Black Isle to view the newly released red kites. There were nine kites over the release area, seven of them feeding, and while we watched them, a common buzzard flew by, and then a pair of honey buzzards, the female carrying wasp comb. To make it extra special, a pair of ospreys were displaying above us and landed in a tall tree. It was great to share that moment with Ian, who had been such a supporter of raptor reintroductions.

In the meantime, Colin Crooke had come up with a potential release area beside Loch Maree in Wester Ross. He had got to know Dougie Russell, the stalker, and also his ghillie, Stewart Keenan, who lived at a remote bothy. Colin knew the area well and on 17 September he took me to meet Dougie and his wife, who were intrigued by our plan. After some exploration, we agreed that the best place for the release cages was on a thinly wooded hillside of oaks and birches, with an incredible overview of the loch and its wooded islands. It reminded both of us very much of places where we had seen sea eagles in Norway. A short distance below the cage site was a big grassy opening, fringed by tall mature larches, which was ideal for feeding the free-flying young eagles with carrion following their release. It was a secure site away from public gaze. Dougie was sure the Dutch landowner, Paul van Vlissingen, would like the idea and allow us to use the area for the project. I wrote to him in October 1992, and received a positive reply, so the Letterewe estate became a key partner in the second phase.

On 26 January 1993 I flew from Aberdeen to Trondheim, via Stavanger, Bergen and Ålesund, and next day had a productive meeting at the Norwegian Nature Conservancy HQ with Morten Ekker, Torgeir Nygård and Alv Ottar Folkestad, to outline the request for further licences, which we later received, and to discuss our new project and the need for more young. Before leaving the building I met Ola Skauge, the Director, for a half-hour discussion about various mutual nature conservation issues, and then Torgeir took me to NINA, the science institute, to talk through logistics for the summer. The next day, Torgeir gave me a birding tour of the local fjords and we saw one adult sea eagle but missed a Barrow's goldeneye that had been causing a stir for the Norwegian birdwatchers. That evening I caught a plane to north Norway and was collected from Bodø airport by Harald Misund. I stayed two days with Harald in a very snowy Norwegian winter scene. We saw several sea eagles, including his local pair, and he agreed to assist the Scottish reintroduction again. This was a very kind offer, as I knew, from field experience, that collecting, housing and feeding young raptors for a project is very time-consuming and onerous. His agreement allowed me to report back home that the project was underway.

Colin Crooke, Greg Mudge of Scottish Natural Heritage and Stewart Keenan built the cages to a design which I had used for our successful red kite reintroduction on the Black Isle. This was a new idea for large eagles, as two or three young would be kept in each cage; human contact would be avoided by having a wooden back to the line of cages. Food for the eaglets would be supplied through a sleeve in the back wall directly onto the nest platform, and each 'brood' of young would not be able to see the eaglets in the other cages, in order to mimic the wild situation as closely as possible. The young, when ready, could be released by lowering the cage front at dawn. On 20 May, when the release site was ready for inspection, I visited it with Laura Gunn, the government veterinarian in Inverness, who passed it ready for use.

On 7 June I met Flight Lieutenant Steve Rooke at RAF Kinloss for the briefing at Squadron 106. I gave a short talk on the project to the Nimrod Captain, Steve Fawcett, and his crew and then we departed. The crew carried out their secret military surveillance duties in the North Sea and off Trondheim, before landing at Bodø at 1 pm. John Love, who had left earlier by scheduled flights, and Harald Misund arrived with 10 eaglets, which were quickly stowed aboard as the captain kept two engines running. We landed back in Moray at 1 pm, where we cleared customs and health inspection. The eaglets were driven to Letterewe, while I called on John Easton, the Conon Bridge veterinarian, who drove with me to the release site to check all the birds. We placed them as two broods of three and two of two in the four cages. The next stage was underway. I was a bit worried that some of them were rather young and had become a bit tame while in captivity in Norway.

Following some discussions when coming home from Norway, Steve Rooke organised an aerial survey of potential nesting areas in the Minch. On 25 June we met Ian Campbell, a Sea King helicopter pilot with 202 Squadron at RAF Lossiemouth. It was a training flight, so while over the Moray Firth they carried out life-saving practice by lowering me on the winch to just above the waves and then pulling me back up. We saw three pods of bottlenose dolphins in the firth and then crossed the mountains to the Minch. Our main search was of the Shiant Islands, where sightings of sea eagles were becoming regular. I scanned all the sea cliffs but found no sign of breeding eagles, so we turned back and checked several sections of the Wester Ross cliffs on the return flight.

Stewart kept the eagles well fed from food sourced by Colin and Dougie, and the veterinarian cleared them from quarantine on 15 July. That day, when we passed Stewart's bothy, a female pine marten walked out of a hole in the grey slate roof of the old barn and lay down in the sun, a large young came out and tried to suckle, then both climbed back into the den. We decided to measure, weigh and tag the young sea eagles on 26 July and were delighted that Paul van Vlissingen and his partner could come and view them while we processed them ready for release. Paul was so enthusiastic about the project and proud that it was being done on Letterewe. We had 10 young to tag, but the onslaught of midges, clegs, sweat flies, mosquitoes and deer keds meant we could only manage one cage at a time, even with repellent and midge candles, before we retreated to Stewart's bothy for a break. To add insult to injury, his Old English sheep-dog had fleas. We would have a quick coffee and then head back to do more before being forced into another break. I don't think there's anywhere in the world worse for biting and irritating insects in July. I pitied Stewart, who had to put up with it every day. We decided to leave the youngest ones until 7 August.

On 12 August we got to Stewart's bothy in the evening and stayed overnight so we could be up at 4 am to open the cages. After a quick coffee we walked up the goat track, as the day dawned, to the viewpoint high above the cages. At precisely 5.43 am, the first eaglet (number 1) flapped out below us. Number 3 left at 6.15 am, number 2 at 6.40 am and the fourth at 7.22 am, all of them settling in the local woods. It had been a successful release, and at 8.10 am we walked down in the rain to have breakfast. The rest were released on 29 August.

Stewart was now in charge. At night he put out carrion fish and meat on big rocks in the clearing, and some by the cages. The young eagles quickly learned the routine and it proved to be an excellent location. Colin, along with Greg Mudge of Scottish Natural Heritage, were regular visitors, and on 22 October it was my turn to feed them at the weekend, while Stewart was away. I drove over in the afternoon and it was raining as I got to the clearing where three eagles were in the big larches. I put out salmon scraps and three bags of venison chunks. Two ravens viewed from high up and a buzzard crossed the clearing – the sea eagles clearly had neighbours taking some of their carrion. As the rain cleared I hiked the goat path to the viewpoint and stayed until dusk. I saw four different eagles and managed to read the coloured wing-tags of two of them. Everything looked good.

One of the eaglets, with red wing-tags and numbered 7, was last at the release site on 4 November and then spent part of the winter, December to February, at Munlochy Bay, where we watched her one evening as she hunted roosting cormorants in the big trees growing on the cliffs. She was seen in the Orkney Islands on 24 and 25 April 1994, and next day was at Sumburgh in Shetland, before returning to Orkney. Amazingly, in late October, she was identified in west Norway and then, in May 1997, she was seen further south in Rogaland. In 2001 she paired up and bred; between 2002 and 2005 the pair reared five young. She then lost her second tag and was no longer identifiable. I was very interested in the return of this erne to Norway, for it supported my old suggestion that at least one, maybe two, of the sea eagles I released at Fair Isle did likewise.

In 1994 I visited Letterewe on 18 January and watched two of them, 600 metres above me, in gale-force winds, wings tucked in, rapidly gliding away past the cages. Kate Thompson took over the care of the eagles and on 15 June the veterinarian and I visited to check out the cages for the new eaglets. I went to Norway on 21 June to stay with Harald. The next day we visited the eyries on various islands and collected four young from different nests containing two. It took several days to collect 10 eaglets and the Nimrod arrived at midday on the 27th. We had a good flight back with Steve and Tom, both members of the RAF Kinloss bird club, and when we arrived at Kinloss, the Duke of Edinburgh was on a visit, so the Commanding Offficer brought him along to view the sea eagles. Greg Mudge, Lorcan O'Toole and I drove them to Letterewe and – the joys of sea eagling – we finally got home at 3 am. John Easton, the Conon Bridge veterinarian who had helped us with many raptor issues over the years, came with me to Loch Maree on 20 July to examine the birds and clear them through quarantine. Colin, Kate and I wing-tagged them on 2 August, when the midges were absolutely appalling. All were released later in the summer and followed the previous year's behaviour, although occasionally the 1993 crop of youngsters joined them at the feeding site. I loved visiting them in winter, so I did the 'feeding trip' on 3 December, another windy and wet day, clearing after midday. Two snow buntings were on the track, and then, from the viewpoint, I saw five of that year's young, including four together very high above me over the cliffs. In the strong winds they were cartwheeling and rolling and, on a few occasions, two would grip talons and spiral down for some metres. They were clearly enjoying life and, for such large birds, were very agile. I lay mesmerised in the heather, looking up at them. I later saw two of them flying with two golden eagles further along the loch. Plenty of food was still at the rocks, so I kept what I had for the deep freeze, and called at dusk at Dougie's cottage for a coffee and a yarn at his fireside, catching up on local news.

From 1995, Kevin Duffy took over the work of the project at Letterewe and continued until the release of the final group in 1998. I first met Kevin at the release site on 1 June when I visited with Alastair MacNab, the government veterinarian who cleared us for another season. Kevin had worked with Mauritius kestrels and I quickly saw that we had an excellent raptor enthusiast to carry out the remainder of the project. I went to Bodø on 20 June and we flew back with six young, collected by Harald. Colin, Lorcan and I drove over on 19 July to check and wing-tag the eaglets with Kevin, and they were released later in the summer. Ten more arrived at RAF Lossiemouth on 20 June 1996, brought over by Harald and John Love. In 1997, 10 young were released after we tagged them on 31 July and 5 August.

* * *

Our last flight in the RAF Nimrods to Bodø was on 18 June 1998. All the flights for the sea eagle project had been organised by Flight Lieutenant Steve Rooke of RAF Kinloss. He was a major player in the sea eagle's return to Scotland, and was also a talented wildlife artist who painted some beautiful raptor pictures. When planning for the last trip in 1998, Steve asked me if there was anywhere I would like to see which could be included in their maritime training flights in the Arctic. I replied that I'd love to see the remote Norwegian island of Jan Mayen, so off we set from Moray one morning, flying north over Sutherland, our last glimpse of the UK the remote island of Sule Stack, where we could see the white guano of the gannet colony far below. Finally, after a long flight, we sighted the incredible island of Jan Mayen at 71 degrees north – the 2,300-metre volcanic cone topped in snow rising from a grey ocean landscape. The plane circled the big island twice at a distance and then set course east for north Norway, passing over miles and miles of pack ice, where during low-level flight, we could see harp seals resting on the ice next to their blow holes. We tried hard to see a polar bear but failed, sadly, even though the rear windows of the Nimrod – which could be unscrewed for photographic purposes at low speed – also gave a great view with binoculars.

We reached Bodø after nearly eight hours' flying. Harald had collected 12 young sea eagles and the next morning they were loaded onto the aircraft for our return flight. On this occasion we had to clear customs and animal health checks at Glasgow airport, where we were joined by the Scottish Natural Heritage Chairman, Magnus Magnusson, for the return to Moray. There we were met by two TV crews as well as news reporters and cameramen. Magnus thanked RAF Kinloss for all their valued support over the years and presented them with a framed portrait of the sea eagles.

Interestingly, 1998 also saw the successful breeding of two pairs from the Letterewe-released young, with three of the breeders just four years old. This input by new pairs allowed the annual number of young reared in Scotland to pass, for the first time, the magic number of 10. That year also saw the first breeding of second-generation Scottish sea eagles, another milestone in the project. The year 2000 was another special one. In Scotland the number of breeding pairs had reached 25 and the hundredth young in total was reared. So when a group of us, including John Love, Roger Broad, Mick Marquiss and Alison MacLennan, attended the Sea Eagle 2000 Conference at Stockholm, we were able to be admitted as one of the countries with breeding sea eagles. Our venue was outside Stockholm in the Swedish archipelago, and on the first day we were taken by boat to view the local sea eagles and their island-studded home. It was a marvellous gathering of eagle researchers and conservationists: I particularly enjoyed talking with Vladimir Masterov about his studies on white-tailed eagles and Steller's sea eagles in Kamchatka, eastern Siberia. He gave me some beautiful photos to use in lectures and an invitation to try to visit him. I noted in my talk, entitled 'An enhanced recovery programme', that there were many 'empty chairs' at our gathering: the countries of southern Europe which no longer hosted the species. I suggested, in fact, that there was too much emphasis on regarding the sea eagle as a northern European species and a failure to recognise that the breeding range should extend south to the North African coast. I was pleased that a resolution, which I drafted, was on the list of conservation actions at the end of the conference: 'the species should be encouraged to return to those countries where the species had been exterminated in earlier times'. I will address this in subsequent chapters.

Later that year, on 30 November, the RSPB and SNH organised a special 25th anniversary party at the Aros Centre near Portree on the Isle of Skye. I drove across with Colin Crooke, who had been through many sea eagle adventures with me. The new sea eagle film was premiered and we had a lovely evening of ernes, music and chat. John Love was presented with a framed sea eagle as the 'Father' of the sea eagles (or, as John likes to call it, 'the eagle with the sunlit eye'; in Gaelic 'Iolaire sùil na grèine'). I also received a beautiful sea eagle photograph by Chris Gomersall, framed and inscribed, while the chairman gave me the title of the 'Grandfather' of the Scottish sea eagles, even though I was just into my 60th year. It hangs on our dining room wall and is a reminder of a wonderful episode in Scottish life.

Field surveys were an important part of the sea eagle project, and staff from the RSPB explored many areas of the north and west Highlands and the islands. It was dedicated fieldwork in remote areas, often alone, covering the ground on foot. My diary records a typical day when Colin Crooke and I spent 12 May 1999 checking sites in Wester Ross. We started near Gruinard and, after walking miles, we had to wait for the rain to cease before viewing a tree eyrie with one young. Our next stop was the Rubha Re headland, where we carried out a long search; at one stage Colin stayed at an excellent viewpoint while I searched the wood below to find a previous nest blown out. I noted many quad bike tracks and evidence of muir-burning, so we concluded that the eagles had moved. As ever, we collected data on other species as well as on land use.

Since then, the sea eagles have gone from strength to strength, the annual totals growing from 25 breeding pairs in 2000, when the 100th chick also flew; by 2006 over 200 young had been reared, and presently the population stands at 140 pairs extending from the Inner Hebrides north to Orkney. The first pair bred successfully on the Orkney island of Hoy in 2018, and it's been marvellous for many of us to see the ernes restored to many of the ancestral cliffs where they used to breed. Some of them have surprised us by breeding at inland locations, such as beside Loch Ness and in Angus and Strathspey, but even there they are associated with the large rivers and freshwaters. These were the result of the east-coast releases in Fife of 85 young from Norway between 2007 and 2012, although they tended to settle in the mountains or even in the west, rather than on the eastern coast and estuaries. Even in the early years, the 1980s and '90s, some sea eagles from the west came inland to scavenge dead and dying salmon in late autumn, after the seasonal spawning, in the big gravelly rivers and tributaries of the River Findhorn and the River Spey, in many ways mirroring the scene two millennia before on the same rivers when the sea eagles were in the company of salmon-eating brown bears and wolves. It's a marvellous thought. Fortunately, one pair settled to breed on Tentsmuir National Nature Reserve in Fife and reared their first eaglet in 2013, but disappointingly it has not been followed by more pairs along the sea cliffs and estuaries from Angus to the Firth of Forth. I hope that recolonisation will occur one day. The fact that the east coast-released birds drifted away to breed with the established population suggests that subsequent releases should occur further away, for example in the Solway Firth and southwards.

Between 2008 and 2010, bringing us bang up to date with present-day technology, we satellite tagged six young sea eagles on the island of Mull with Dave Sexton of the RSPB. The Mull and Iona Community Trust and SNH funded the purchase of the transmitters. On 18 June 2008 we tagged the first two young sea eagles at nests in Forestry Commission forests, and the television filmmaker Gordon Buchanan came along to record the day's fieldwork. The two eaglets were named Breagha and Mara. On 26 June 2009 we placed tags on two more young, named Midge and Oran, and on 23 June 2010 I was back on Mull with Dave Sexton and colleagues to tag two more young, which were named Shelley and Venus. This proved to be a very exciting project, with the Mull Eagle Watch Project able to show how the young from the Isle of Mull explored Scotland. Their exploits were eagerly followed by people from all over the globe via the internet and social media. We were thrilled when the males Mara and Midge settled down with mates to breed in Ardnamurchan, while the female Shelley, who spent much of her time in Lewis and Harris, was located breeding on a coastal cliff in northwest Sutherland. Sadly, the other three all disappeared off our screens, almost certainly killed illegally. Oran was last located by GPS transmissions south of Oban, Breagha in the Mallaig area and the wanderer Venus in the Cabrach hills.

When I look back, I'm disappointed that so many young sea eagles have undoubtedly been killed illegally by man, either by shepherds and crofters in the west and on the islands, or by gamekeepers on the red grouse moors of the east. Some of the early casualties probably died accidentally when poisoned baits were set out to kill crows and ravens, albeit illegally. There have been some complaints about the eagles killing lambs in some areas, although the great majority of pairs have caused no such concerns. There's no doubt that dead lambs, and even full-grown sheep carcasses, have been eaten by sea eagles, as well as by golden eagles, ravens, crows and gulls. There have been some cases where lambs have been taken by individual sea eagles, but scientific studies have failed to identify serious losses, while post mortems have shown that many of those lambs had been unlikely to survive for health reasons or were dead before being scavenged. Bad weather at lambing time is the biggest killer. Sadly, the arguments over the sea eagle's return have not so far been resolved, even by funding and management from the Scottish government. The problem with this approach, throughout the world, is the argument about whether or not the predator killed the animal; and then the clash, often very emotional, between locals and authorities.

The most encouraging thing for me is the tremendous joy that people have felt, and still do, on seeing their first sea eagle in the wild in Scotland. Even better, it's now possible to go in a boat with a guide and watch them coming to take fish from the sea – it's a dramatic chance to see the great bird at close quarters and to take fantastic photographs. Sea eagle watching has created eco-tourism businesses that have attracted up to £5 million a year to Mull and £3 million a year to Skye. Several

Adult male sea eagle with fish on the Isle of Skye in August 2017. The dramatic MacLeod's Maidens in the background were ancient breeding areas for Scottish sea eagles.

times I nearly managed – but failed – to accompany Colin Crooke on trips with a Portree boat owner who, in the early 1990s, pioneered showing people white-tailed eagles swooping down to take fish. The years went by, but finally, in August 2017, my friend Mike Crutch invited me to join him on a day trip to Skye to photograph sea eagles, heading out from Carbost pier to the great cliffs of Skye, where they regularly nest – in 2017, this pair of sea eagles had an exceptional brood of three flying young. The eagles knew what was going to happen, for as the boatman, Steve Hopper, threw a fish well away from the side, the male plunged from the cliffs and, in a sweep of huge wings, grabbed the food from the water to the noise of camera shutters, just yards from us.

What a fantastic sight – and what a remarkable story. The great ernes are back, breeding again in Scotland, and our land and seas have regained some of the wildness of earlier millennia. Once it was sea eagles with Neolithic people and their stone tools, and now they are with us and our 21st-century ways.

A male goldeneye duck in Moray. (*Photo by Gordon Biggs*)

2

# GROUNDWORK

## *Learning my trade*

In April 1960 I had moved to live and work at Loch Garten, south of Inverness, having joined the RSPB's Operation Osprey team, protecting the only pair of nesting ospreys in Britain. Where they bred in the ancient Caledonian forest of Abernethy, though, there were also other superb new birds for me, like capercaillie, Scottish crossbill and crested tit. There, I was very fortunate to have two special mentors: George Waterston who, while at Fair Isle the previous September, had encouraged me to apply for the post with the RSPB, and Wing Commander Dick Fursman, recently retired from the RAF, who was the warden in charge. Both encouraged me and gave me much valuable advice in my early twenties. In those early days, I loved protecting the breeding ospreys and showing them to the many visitors to our observation hides, but with the Cairngorm mountains and much of the Highlands as close neighbours, I also explored and gained a huge amount of experience, not just of fieldwork in the Highlands but of how to work with people from all walks of life. I made many lifelong friends and built a reputation as a field ornithologist. I was always trying new things, finding and studying all sorts of rare breeding birds and

Famous Loch Garten osprey nest tree in the early 1960s when the author was an RSPB warden. This was a spring visit to fix an electronic alarm system on the trunk.

enthusiastically embracing a life that was never going to be 9 to 5. In retrospect, that was the groundwork for the rest of my working life. I also learned much about ospreys and how to conserve them, of course, but more of that later. Let's start with two other birds special to me: the goldeneye and the peregrine falcon.

## Encouraging goldeneye ducks to breed in Scotland

Goldeneye ducks have been a favourite of mine ever since that first spring in Scotland, in 1960. They were to be found as a winter visitor from Scandinavia on many of the lochs in Strathspey. One of my favourite birdwatching places that first April was a short walk from our tented camp in the forest, through the ancient pinewoods to the edge of Loch Garten. Pushing my way, soon after dawn, through heather and blaeberry, I would often flush huge capercaillies and could listen to the trill of crested tits. Reaching the edge of the loch, I would sit down with my back against a 200-year-old Scots pine, sheltered by its spreading branches. On still mornings the loch would be flat calm, the trees on the other side mirrored in the dark water with maybe some mist being lit by the sun rising behind me over the forest. Immediately, I would see small groups of goldeneye on the water, often in full display. The males are brilliantly patterned black and white, the head black with a green sheen and brilliant orange eyes. They would be throwing back their heads and displaying to the females with an eerie, creaking call and making short rushes of splashing water towards them. Further over might be half a dozen red-breasted mergansers, also indulging in noisy, boisterous displays. Their females would nest somewhere close by, but by early May all the goldeneye would suddenly disappear and fly to their breeding grounds in Scandinavia. Most lochs held these over-wintering goldeneye;

on 27 April 1960, for example, I saw 46, including 17 drakes in adult plumage, on Loch Morlich. Some pairs lingered until mid-May before heading to nest in Sweden, but I always hoped that, one day, they might breed with us.

The goldeneye is a small diving duck, widely distributed through the boreal regions from Scandinavia east to Siberia and migrating to more southern latitudes in winter. They dive to search for food – invertebrates, molluscs and crustaceans – in fresh and salt water. They can be found in large flocks or in small numbers on remote lakes and rivers. The goldeneye is a tree-nesting duck, and it chooses to lay its clutch of eggs in holes of trees, including, in Scandinavia, in the holes bored for nest chambers by the large black woodpecker. The female encourages her brood to jump out of their tree nests and escorts them to fresh water, where they catch their own live food. There are no historic breeding records of the species in the British Isles, but it may have bred there in earlier millennia.

On 12 July 1960, Charlie Cowper and I were looking for hen harriers on Dava Moor, about 20 kilometres north of our famous osprey eyrie at Loch Garten. We explored Lochindorb with its ancient island castle and then some small lochs at the north end of the moors. At the lovely Loch Allan, which lies nestled in moorland, we were watching a small group of tufted ducks when I suddenly saw a female goldeneye furtively paddling along the far shore with, scurrying behind it, a tiny duckling. I was very excited as I thought we had found the first breeding evidence for the species in Scotland and the British Isles. But, alas, the duckling was the wrong colour; it was dark brown, like a young tufted duck, rather than black and white, and I guessed that the goldeneye had adopted a lost tufted duckling. Nevertheless, this was enough to start my heart racing and I wrote that evening with my news to George Waterston. He checked the books in the SOC library in Edinburgh and confirmed that the duckling was not a goldeneye, but thought it was exciting enough to try putting up a nest box at Loch Allan. George and I went together to the loch a fortnight later, and the female was still present, but we saw no sign of the duckling.

That winter, George arranged for the Blind School in Edinburgh to build 24 big nest boxes, and he wrote inviting me to meet him on 2 March in Edinburgh, to start the Operation Osprey season. The boxes were sent to Strathspey and, having sought and easily received permission from landowners to carry out the project, I started to put them up beside suitable lochs. My diary entry of 16 March 1961 says that I borrowed a canoe from Dick Fursman and paddled out across Loch Mallachie, near Loch Garten, to the small island at the southwest end. There were lots of signs of otters and an old mallard nest. I wired a box into position in the fork of a balsam poplar, facing northeast, about 5.5 metres above the water. It's a good loch for goldeneye, and a pair stayed very late into spring. Later a pair of starlings took over the box and reared young. The next one I put on a tree at the south end of Loch Garten on 19 March, and next day I was back at Loch Allan. I waded out to a small island at the north end of the loch and wired the box, again about 5.5 metres up, in a dead rowan tree leaning over the water, facing northeast. I took it as a good sign that two male and three female goldeneyes were present. On the following days I erected nest boxes at various lochs in Badenoch and Strathspey, including a very nice location at the southwest end of Loch an Eilean. During the fieldwork I also found some old duck boxes, for goosanders, which I later discovered had been put up in the early 1950s by a man so keen on encouraging birds that he went by the name of 'Nestbox McKenzie'. They needed cleaning out and I fitted new lids.

Goldeneye nest box erected by the author in Strathspey in 1961;
on a Scots pine overlooking a freshwater loch.

At the end of the season I went round the boxes to check them and, while I found no evidence of goldeneyes, saw that several had been used by starlings and one by redstarts. I regularly kept an eye on the boxes throughout the next couple of years, but no goldeneye were tempted to use my boxes; instead, I found them occupied by tawny owl, jackdaw and goosander, as well as starling and redstart. In autumn 1963 I became the warden at the Fair Isle Bird Observatory in the Shetland Isles, but when I was off the island in the winters I still made regular visits to my boxes to secure them and top them up with sawdust; strangely, I found that starlings were extremely meticulous about removing sawdust from the boxes before they built their own nests. Without a sawdust lining, though, the boxes were unsuitable for goldeneyes. I noted that someone had moved some of the boxes from the edge of the water into woodlands, presumably to encourage tawny owls. Sadly, the project did not seem to be working, but I was determined to persevere and succeed. Tenacity and a long view to the future are important in wildlife conservation.

Wintertime in the Highlands was a good season to look for goldeneye among the wildfowl flocks, both inland and on the coast. In the early 1960s numbers were relatively small, with a hundred being a good count on the coast, but after the very severe cold winter of 1962–3, when more wildfowl came over from mainland Europe, numbers increased dramatically. There was plenty of food for birds to find at the sewage outfalls, used also by distilleries and breweries, with whisky and beer production booming. The biggest outfall in the north was at Invergordon, where a grain whisky distillery was built in 1961, with a major increase in production in 1963. The effluent pipe was only a couple of hundred metres off the front of Invergordon in the Cromarty Firth. My most amazing visit there was on 15 January 1966, when my notebook recalls a major feeding flock around the end of the pipe: '950 goldeneyes along with 455 mute swans, 391 whooper swans, 8500 wigeon, 100 mallard, 170 pintail, 256 scaup, 50 tufted duck and 50 pochard – all hungrily feeding on the spent grain.' I was standing there counting when a local turned up, who knew me as the 'birdman' who caught and ringed swans. He told me I should have been there the week before, when I could have ringed the lot: there had been an alcohol release problem at the distillery and the swans, he said, had all been wandering, drunk, along the beach. The Longman at Inverness was another favourite place, offering a mix of sewage and distillery effluent – my highest counts were 450 goldeneyes on 11 January 1966 and 596 on 16 November 1973. My third locality was Burghead, on the Moray coast, where a whisky malting plant was built in 1966 and the outfall attracted up to 500 goldeneyes. These counts were exceptional but not as high as the several thousand goldeneyes attracted to the brewery effluent off Edinburgh. I think these high numbers may have helped stimulate the breeding colonisation. In the 1980s, tighter effluent controls meant that far less food was pumped into the sea, and duck numbers fell. Even on inland lochs the numbers of wintering goldeneye were much lower than they had been in the 1960s and '70s. Further recent decreases are probably due to milder winters further north. In fact, on 3 December 2017, the coordinated waterfowl and wader count of the Moray Firth coast from Brora to Buckie found only 112 goldeneyes, which is a massive decrease in numbers.

1970 was my last year at Fair Isle, and the great excitement in Strathspey was that in mid-July a visiting birdwatcher, Mr CP Mapletoft, had found a female goldeneye with four ducklings at a small loch called Lochan Mhor, in Rothiemurchus. On 23 July they were also found by Tim Milsom, who reported the sighting to the local bird recorder, Doug Weir. Doug saw the four young still there on the 29th. He had seen a female on that loch in early May and also a drake until late June, so he thought they had nested somewhere nearby. There had been occasional sightings of birds summering from 1967.

In January 1971 I started my new job with the RSPB as Speyside Representative, later as Highland Officer, and my family first rented a house in Boat of Garten. I had the task of being the RSPB's man on the ground in northern Scotland – Strathspey in particular – with a remit to carry out the better conservation and protection of birds, one of which was clearly the goldeneye. My diary entry for 8 February 1971 shows that I had started to check the boxes that I'd put up all those years before. It was a fine, clear day as I walked down through the ancient forest to the eastern side of Loch Garten and checked the two boxes at the end of the loch. The one on the dead tree needed a lid and there was some debris of dead reeds in the bottom, while the bottom of the nest box in the live Scots pine was faulty and it also needed a lid, so I clearly had some work to do in the next month to make certain that all the boxes were up to scratch and lined with sawdust. On 17 March I visited Loch Mallachie, where two boxes were okay, but two others had missing lids. On the loch were five male goldeneyes and one female, and I found no breeding goldeneye in 1971, despite visiting the location where they had bred the year before.

By October, we were living at the Old Manse in Rothiemurchus, with the 1970 goldeneye loch not far behind our home. On 12 October I walked to Loch Gamnha in the morning to check old nest boxes. It was a cold autumn day with snow showers falling as I hiked along the shore of Loch an Eilean, looking across at the ancient castle where, 70 years before, ospreys had nested. I found the old nest box, and as I climbed the pine tree, six red deer stags were roaring on the hillside above me and two flocks of greylag geese flew overhead to the south. The lid was missing and nothing had used it. On the way back I checked another of my original nest boxes at the far end of Loch an Eilean. When I lifted the lid, as I perched high in the Scots pine, I was very excited to find evidence that a female goldeneye had nested there in the summer. There were two damaged eggs in the nest, partly incubated, and also broken eggshells, down and feathers on top of sawdust. The intact egg measured 59.19 x 42.55 mm, and the down was very pale greyish with lighter centres, even a slight touch of blue; the small feathers were white. The box was well scratched inside. I was certain a goldeneye had fledged young that summer and that the ducklings must have been predated when she was moving them, because we never saw them. A careful examination of the nest material convinced me that this nest box was where the female goldeneye had laid her eggs in 1970 as well, so the very first breeding had occurred in one of the boxes I had erected in the early 1960s. What a long time to wait for success, but definitely well worthwhile. Later, at home, I fitted a new lid to the damaged box and fixed it in a birch tree at nearby Loch an Mhor, a superb small loch in the woods. While there, I climbed to check the other two boxes. I had a feeling that breeding goldeneye were going to become an exciting part of my year.

On Sunday 28 May 1972, Colonel Iain Grant, the Laird of the Rothiemurchus estate and a great friend, arrived breathlessly at my house in the afternoon with news of young goldeneye at Loch an Mhor. I immediately went back with him and found a female with nine tiny ducklings, newly hatched. They were dark brown with white underparts and chins, and white patches on the wing stumps. They were very active, diving and feeding on insects from the surface of the water, and also present was a drake in eclipse plumage, possibly the father, and an extra female, plus six male and four female tufted ducks. In the evening I went again, watching young long-tailed tits on the way, and on the loch a female wigeon with eight ducklings and little grebes with their young. I checked the nest boxes and the first one in the bay held a beautiful goldeneye nest, well lined with much down and broken eggshells, while a goldeneye had also been in the second box. That, though, was now home to a young tawny owl.

Next day was cold and wet with strong winds. I went to the loch in the afternoon and saw six goldeneye chicks feeding across the south end of the loch, but they were very difficult to count as they were busy diving and feeding. Then the female arrived by air, swam to the end of the loch and called: seven young came to her and, a few minutes later, two more, to make the full brood of nine. I was surprised she had been absent and wondered if she had been checking where to move them. The young had dark bills, dark fronts, two white stripes on their sides and two white spots on the rump with whitish underparts. They were the most delightful ducklings and really independent. I checked again in the evening of 2 June, with my elder son, Gavin; a pair of crested tits with at least two young were at the north end, and the female goldeneye still had all nine young, fit and well. A pair of coot now had chicks as well.

A brood of newly hatched ducklings on Loch Insh.

When I went to the lochan on 4 and 5 June, the ducks were gone. It was clear that they had departed or been predated, but three days later, at Loch Alvie, I found nine young goldeneyes, all looking well. It was the same brood, and proof that golden-eye ducks move their young quite long distances – the mother duck had walked or swum them across two roads, the main railway line and the River Spey, a distance of over 4 kilometres. There was also a male and two females, and eventually one of the females, at last, came and took the chicks in tow. A large fish, probably a pike, chased them at one point. On 25 June Gavin and I were in Aviemore for the Sunday papers and had a quick look at Loch Alvie; seven of the young goldeneyes had survived and were now half-grown, the female with them. The 1972 season was a big success after the failure of the previous year.

1973 was a year to get more boxes in order, so on 7 April, with a cold northwest wind and snow showers, Gavin and I went to Loch Pityoulish. We walked down to the bay and erected two boxes on the Pityoulish ground, one a little over a metre up in an overhanging larch, the other higher in a birch tree. Further west, I refound one of my 1960 nest boxes on the Rothiemurchus estate in a large birch: it contained a clutch of 10 goosander eggs from 1972. I cracked open three of them and found that they had been incubated the previous summer – maybe the female had been predated when off the nest. I removed the eggs and put in fresh sawdust; three goldeneyes were on the loch. On 12 April I talked with the Kinrara estate keeper, Mr Polson, and fixed a goldeneye box in a birch tree on the Loch Alvie peninsula. Later in the afternoon, I visited the Bogach nearby and put up two more nest boxes. This beautiful hidden loch held an adult male and two female goldeneyes, four pairs of tufted duck and a pair of teal, as well as a big colony of black-headed gulls. When I checked these boxes on 17 April there were two male and four female goldeneyes displaying near the boxes, and one female had a brighter-coloured eye and bill, which I later learned was a good sign of a potential breeder. That evening, I took my boat out to the island on Loch Insh, mended one goldeneye box and removed an old jackdaw's nest from it, then put up a new box at the west end of the island.

The first ducklings of the year were seen on 12 May, so I went to Loch an Mhor at 7 am two days later. I climbed to the first box and saw that it had been newly used, with a beautiful nest of down; then, below me, I saw eight tiny ducklings paddling out of the bay. I watched from the far end and the ducklings moved to the other end, where they busily fed along the edge of the sedge beds. The mother flew in low from the other pools and landed with the young; two goldeneyes, which were already on the water, swam towards her and she chased them away. The older one was very persistent and it called for much chasing underwater and lying along the water in a threatening posture to keep it away from the brood. It was superb to watch. I checked the second box but found nothing in it. The loch also held three pairs of incubating coots, four pairs of little grebes, a pair of wigeon and three pairs of tufted ducks. After lunch I returned to the loch, finding the female on the water near the box, but I couldn't see young from the far end. Briefly she chased off a red squirrel that was on the water's edge. On 17 May I made an evening visit to Loch Alvie and found a new female goldeneye and six ducklings on Loch Beag, the young feeding busily along the edge of the sedge beds.

On 28 May I walked with Gavin and his younger brother, Roddy, to Loch an Mhor. We first saw a female goldeneye with eight young ducklings feeding off the nest box bay, then another female, with six larger ducklings, in the east bay. After tea I walked back to the loch and climbed to nest box 4: tawny owls had been in it and probably reared young. Then I went up to nest box 3, which had been used by goldeneye, and where some hatched shells remained. It was excellent to see both goldeneye families present. On 6 June I returned to the loch with Brian and Linda Marshall, friends from Shetland; two broods of goldeneye were present, one of seven ducklings without the mother and an older brood of six ducklings with a female. A week later there were still at least four large young and seven smaller ones, making three successful broods in 1973. In complete contrast, on 30 November, I walked on thick ice to the island on Loch Mallachie; I had no luck there, although one box had been used by starlings.

Nineteen-seventy-four started with more box work. On 12 March, for example, I drove out to Loch Ness to see Lord Burton and 10 days later I put up two boxes on his loch: one on the north shore side, 4.5 metres up in an alder, the other one halfway along the loch, 6 metres up in a silver birch. Four goldeneyes were on the water. In the evening I put up a goldeneye box on a small loch behind our house. On 19 May I found a brood of nine newly hatched ducklings on Loch Alvie with the female protecting them from an adult male and three immatures. Our summer tally for 1974, then, was three broods, the same as the previous year.

Nineteen-seventy-five saw me increasing the numbers of nest boxes and spring-cleaning the rest. On 29 March I went to Loch Mallachie, where 20 waxwings from Scandinavia were whirring through the trees over the loch; that was another species we hoped would breed with us. I paddled out to the islands; at the nearest, the two boxes were empty and, on the other island, one had an old starling nest and one was empty. There were six goldeneyes on the loch and many signs of otters on the islands. On 14 April I drove to Achilty Forest near Strathpeffer and met Tim Damfrey, the Forestry Commission forester. We waded the river to the island on Loch na Croic and put up three nest boxes, two of them at the west end. There were six goldeneyes on the loch, and otters had been very active. We returned to the forest office and I was given a spare roll of steel telephone cable, which we had found excellent for attaching boxes to trees.

The 21st of May was a sunny day with clear skies, and I was out birding at 5.30 am. My first visit was to Loch an Mhor, where, as I remarked in my field diary, 'I found a brood of 10 ducklings on the loch, a few days old, the female was with them some of the time and a young female tried to take them over. Six females on the loch altogether and three pairs of tufted ducks. I walked round the loch; first box had a very thick bed of down and was certainly the nest of the brood. Up in the birch wood a female came off her nest so I climbed up to find five eggs in a well-downed nest. The female flew to the loch calling and I walked down to the loch edge and watched. She was eager to get back so I walked on to Loch an Eilean.' Later, at Loch Alvie, there was a female being very noisy but I couldn't find her nest site. So, in 1975, two pairs hatched broods and a third female probably nested.

On 2 April 1976 I was with Stuart Taylor, the warden at the RSPB osprey camp, checking the goldeneye boxes at Loch Alvie and the Bogach; we then went to the Boathouse where we waded across the river to the island and removed the goldeneye box, which had not been used. So on to Loch Pityoulish, to renovate the boxes there, putting in fresh sawdust and erecting one new box. We removed two addled eggs from the first box. In the evening I checked the goldeneye box in Garten Wood, which held a red squirrel's drey. The next day I went to Lochs Pityoulish and Morlich, walked to the nest boxes and replaced one lid, but none had been used. In the summer at least five pairs hatched a total of 46 ducklings.

By 1977 the goldeneye project was becoming too much for me personally to check every nest, with my time, increasingly, being claimed by other important bird species as well as by our conservation work, especially with the advent of North Sea oil exploration. I then had the enjoyment of showing others the nest boxes and introducing them to my local friends with boxes on their land, so that they could take on a lot of the monitoring. Some were special roving summer wardens attached to my RSPB office, and all of them were excellent birders. In 1977, at least seven females laid eggs, but it was a poor season because of cold weather and only two broods of young were reported.

In 1978 a further expansion of nest boxes was planned by my assistant Roger Broad and myself. On 27 March we went with my fellow conservationist John Lister-Kaye to meet Andrew Matheson, the owner of the Brahan estate, and put up three boxes along the River Conon: it looked a good place, with two pairs of goldeneyes present. On 30 April I went to Loch an Mhor and checked the boxes. Number 19 had been used but had no eggs, while in box 13 the female came off 19 eggs – the top ones were warm. Box 14 had a tawny owl with two eggs, while box 15 had 12 unincubated eggs. On 30 May Hilary Dow, a Newcastle University graduate who wished to carry out PhD research on goldeneye, came to visit. We checked Loch Pityoulish, Loch Morlich and Loch an Mhor, where one nest had been robbed, then went on to Loch Alvie and the Bogach, and so to Lochs Vaa and Dallas and, in the evening, with local birder Dave Pierce to the River Spey at the Doune where 10 ducklings had just hatched. 1978 was a good year, with 12 females laying eggs, and although there were some failures due to a cold spring, about six broods of young were seen and breeding pairs were spreading out. Two clutches of eggs were robbed by egg collectors, which really made me mad. On 19 December we were at

Fortrose Academy to collect 40 excellent nest boxes made by the schoolchildren and did a story for the *Press & Journal* newspaper. In the winter I wrote and distributed *Goldeneye Duck Newsletter No. 1* to all the people who had been involved in or helped the project to that point. It became an annual report until 1990, when I left the RSPB.

It was snowy and cold on 7 February 1979 when Roger Broad and I went to Loch Pityoulish, so we walked on thick ice right across the loch and put three boxes along the loch shore. We were very pleased with the locations but it was quite a long job. We drove via Aviemore to Loch Vaa and again walked over ice to the island and put up two nest boxes. I found the remains of one of my 1961 boxes, which I removed; the other had disappeared. We had good views of a green woodpecker flying through the birches as we walked back. Later, I erected four nest boxes on the River Aigas at Carnoch near Beauly – in fact, 35 more boxes were erected during the winter, and in 1979 goldeneyes laid in 21 sites. Ten of the nest boxes used in 1979 had been erected in the autumn of 1978 (1) or the spring of 1979 (9) and it was very interesting to see how quickly some boxes can be occupied; for example, box 73 was put up on 17 April and was occupied by a laying female on 1 May. Fourteen females were successful and hatched 110 young, notable increases on 1978. Two broods of ducklings appeared from unknown nests in wild locations. One female was killed when flying to feed during incubation, most likely by a peregrine, as a goldeneye leg was found in the nearby peregrine nest on 16 June. As usual, fledging success was very difficult to gauge because the ducks moved their broods so often from loch to river. As in previous years, though, a few broods were recognisable and a survival rate of about 50 per cent appeared reasonable, with very roughly 60 young surviving to half-grown.

Nineteen-eighty and eighty-one were good seasons for goldeneye, and successful brood size was 9.70 and 9.86 young per female as against 7.85 in 1979. The population continued to increase from 14 broods of 110 ducklings in 1979 to 17 of 165 ducklings in 1980 and a further dramatic increase to 29 broods and 286 ducklings in 1981. The first young were on the water by 18 May in 1980 and by 10 May in 1981. Some very large broods hatched out in 1981 – 17 young, 16 young (2 broods) and 15 young – and these were the result of more than one duck laying in a box. As in past years, there was a lot of dumping of eggs in boxes with no subsequent incubation, which we were certain was due to young females. The fieldwork involved in checking all the lochs and keeping nest boxes in good condition was becoming more time-consuming, and most of it was carried out by the RSPB Strathspey roving warden and the local reserve wardens.

Breeding success of goldeneye ducks in Badenoch and Strathspey

| Year | Broods | Ducklings | Average brood |
|---|---|---|---|
| 1979 | 14 | 110 | 7.85 |
| 1980 | 17 | 165 | 9.70 |
| 1981 | 29 | 286 | 9.86 |
| 1982 | 24 | 208+ | 8.67 |
| 1983 | 24 | 209+ | 8.70 |
| 1984 | 33 | 311+ | 9.42 |
| 1985 | 40 | 336+ | 8.40 |
| 1986 | 46 | 390+ | 8.48 |
| 1987 | 37 | 332 | 8.97 |
| 1988 | 44 | 427+ | 9.72 |
| 1989 | 55 | 460+ | 8.37 |
| 1990 | 53 | 529+ | 9.98 |

It was so exciting to find goldeneye breeding in our boxes and the population growing, and the main thrust of our conservation work was to get more and more boxes up in the countryside and to move more and more of them outside Strathspey, so that we could extend the population. We became very good at building boxes and at refining the design. In the end, we even did away with the lid, as we found it was much better to flash roofing felt onto the top as water-proofing, removing the risk of nest boxes losing their lids.

It was very interesting to observe that when we put up new boxes in the main study area in Strathspey, female goldeneye were immediately able to find them and use them. I thought this was probably due to the ducks looking for places where predators were least likely to find them. Some people wondered why we were putting so much effort into building boxes for goldeneye: why couldn't they just find their own nest holes? It was actually a problem caused by man, due to the massive reduction of woodland in our country over centuries, even millennia, and the tendency to remove all dead trees for firewood when foresters were harvesting. In fact, goldeneye could hardly find any natural nest holes. I remember one summer walking five miles along the River Spey, inspecting trees and trying to find natural holes. The main tree was alder, which grows alongside water, and on my whole route I found only one good natural hole. Female goldeneye occasionally land on house chimneys – in North America they are sometimes called the 'stovepipe duck' – and twice people found goldeneye flying round in their homes, while one was found dead in a deserted cottage. Of course, in Scandinavia, as we've seen, the bird benefits from the activity of black woodpeckers, the largest European woodpecker, which is absent from Britain. This species creates ideal nesting sites for goldeneye, often many metres from the ground. Birds that lay their eggs in woodpecker holes often use them the following year, after the woodpeckers. In our country, in the native pinewoods of Strathspey, swifts use the holes made by great-spotted woodpeckers; they keep a careful check on the woodpeckers' nests and, as soon as the young have flown, move in and lay their own eggs. It's actually a wise move, because the eggs of the parasites from the woodpeckers will not hatch until the following year.

Long ago in Sweden, before the advent of timber lorries, it was a difficult life for the early woodsmen, who felled the great trees of the Swedish forests and floated the trunks down the rivers when the ice thawed in the spring. To make it slightly easier, they worked out a way of getting a fresh supply of eggs as they rafted down with their harvests of timber. They erected nest boxes in the trees around the cabins where they spent their nights. The boxes were used by goldeneyes and goosanders, providing the woodsmen with a fresh supply of eggs to harvest at each of their overnight stops. Many years later, naturalists and hunters carried on the tradition, with the goldeneye becoming very used to searching for nest boxes in lieu of natural holes: it's this behaviour which contributed to the species nesting in the Scottish Highlands.

So, we refined our boxes, we eagerly spread them into more and more areas and all sorts of people helped us – individuals, school children and the Forestry Commission all built boxes, often supplying their own wood. Soon we were able to give out boxes to groups and individuals and extend our activities into all sorts of new areas. In those days, of course, we couldn't just save the design on a computer file and email it. Instead we had to type it out, painstakingly duplicate our design sheet and send it off by post. More and more people and groups started their own nest box schemes, even as far south as Kielder Water in

Collecting goldeneye boxes made by boys at Raddery School on the Black Isle. From left to right, Colin Crooke, author and Tony Hinde of the Forestry Commission.

Northumberland. By 1986 we had erected 400 boxes and the Forestry Commission had started to distribute an additional 500 for use in their forests.

A very important feature of our project was that we got to know so many helpful and interested landowners, keepers, farmers, foresters and people who lived along the rivers or beside lochs. They just loved having goldeneye nesting with them and, through this delightful species, we built up many long-lasting conservation friendships.

* * *

I always found it exciting to climb up to a goldeneye nest box during the breeding season and lift the lid to see what was inside. Sometimes the female might be incubating the eggs and would lie very still in her downy nest, maybe making a low hiss, her yellow-fringed black eyes glowering up at me. I would quickly replace the lid and climb back down the tree and leave her alone, to return another time to count the eggs. At other times the female might be away feeding and, before leaving her clutch, would have pulled the down and feathers over the top of them to keep them warm. The eggs are a lovely greeny-blue and I would reach down to count them; they were often gorgeously warm. The normal clutch was between 9 and 11, but sometimes there were far more. We quickly learned that goldeneye do a thing called 'egg dumping', whereby immature females lay eggs in an older duck's box. The two may even be related – is it a daughter doing the laying? – but we just don't know. We would find clutches of up to 19 or 20 eggs and, in exceptional nests, found 22, 23, 26 and 28 eggs, where two extra females had added eggs. The immature females would often lay eggs in their own boxes but not incubate them. We would find smaller clutches of 6 or 7 eggs with – the tell-tale sign – no down surrounding the eggs. The mature females, once they are close to completing their clutches, start to pull down feathers from the breast and belly to create a really warm lining for the nest, whereas the young birds, which are not going to incubate, do not. We kept a record for each nest box of how many eggs were laid, and how many hatched. It was common for a goldeneye to hatch her own and some of the dumped eggs, so we recorded broods of up to 17 ducklings. Often, during my visits, there was nothing in the box, just the bed of sawdust; at other times, there might be young tawny owls or jackdaws, and in the early years a clutch of beautiful blue starling eggs. The worst thing was to open the box and find a wasps' nest hanging from the underside of the lid, which would lead to a very rapid retreat down the tree.

Goldeneye eggs take about a month to hatch, and the young all break out of their eggs at about the same time. They spend over a day drying out and getting ready to leave; and then, during daytime, the mother flies out and down to the ground. She starts calling to them and, with a soft growling noise, encouraging them to come out. The tiny ducklings throw themselves out of the box and float down to the ground; some of them might come down six metres or more. They never come to harm and she controls them tightly in a little crèche on the ground. When, from the lack of piping noise from the box, she knows the nest is empty, she is ready to go. She leads them on a march, the ducklings following as closely as if they were stuck to her tail. I remember one time being really fortunate to be standing at our kitchen window, looking across the meadow to a nest box on a birch tree on the croft, and watching the female calling her young to leave the nest. It was incredible to see the tiny ducklings leaping into space and I was impressed at her ability to keep them together on the ground. The females know where they are, they regularly fly back and forth to the water to feed, but this is a different kind of journey, walking their brood to the loch. Those that nest beside the river, a favourite choice for goldeneye, just plop down and within minutes are on the water, rapidly starting to search for insects. Others, though, may be several kilometres away from water and have to walk through meadows, woods, even follow tracks through the heather and sometimes use burns so that they can swim the final leg to the nursery areas. This is a hazardous time, for they are easy prey for the likes of foxes or buzzards, and when crossing roads a passing car might disrupt the brood. In June 1995 I picked up two lost young ducklings on the road near our home. Later that day I walked to Loch Mallachie and put them on the water, where they were gathered up by two female goldeneyes. They were lucky.

Early in the project, we realised the females were sometimes not happy with the initial nursery area and would set off on another journey with their brood to a new location. Maybe those times when we saw a brood with no female in attendance, which was not unusual, were when the mother was away scouting for a better place. As the years went on we found that their favourite choice for the latter part of duckling rearing was on the River Spey. When Hilary Dow was with us in 1978, before carrying out her PhD in Sweden (where there was much more available data for her to research), she suggested to me an idea of how the broods could be followed. A researcher at Aberdeen University had pioneered a novel technique borrowed from the poultry industry, whereby he injected shelduck eggs at the point of hatching with different coloured vegetable dyes. Red, blue, green and yellow ducklings later appeared on the Ythan estuary, and he could track their progress and see which ones joined crèches and for how long. I thought I had better just check with my conservation director, and I remember the discussion. John Parslow thought it a rather neat idea but then said it was 'best to forget we had this conversation!' Hilary instead put tiny spots of colour dye on the ducklings' cheeks after hatching and was able to show how the mothers moved their broods. Some stayed at the nursery loch and others made longer journeys, even to and from several sites to find the best food for their young.

At the end of the year, when we were clearing out the boxes for the next season, we would check the failed eggs. It was a smelly job, breaking them, but we could tell whether they had been infertile or whether a chick had developed and then died. Occasionally females might be killed when they were away from the nest, as must have been the case when the leg of a female goldeneye was found in a nearby peregrine eyrie. We sent some of the addled and deserted eggs to the Department of Agriculture and Fisheries laboratory at East Craigs, near Edinburgh, for chemical analyses for pesticides, and I was most grateful to Douglas Ruthven for sending us information. A total of 15 eggs (1977, 1978 and 1979) were analysed and the levels were similar to those found in eggs from Sweden and were not likely to cause egg failures. Subsequent tests on more failed and unincubated eggs gave similar results.

Once we felt confident that they would not desert their nests, we started to catch and ring female goldeneye at the boxes, using a large butterfly-type net. With Dave Pierce, I ringed eight females in 1980 and a further 19 in 1981. Four birds ringed in 1980 were re-trapped in 1981, two were in the same boxes, and two others had moved 3 and 2.5 kilometres respectively. All had hatched and fledged young in both years. Subsequently more were ringed and re-trapped in following years, up to eight years and 5 kilometres away. One was recovered in Argyllshire, and we think that the drakes and non-breeders frequented the Beauly Firth and the Clyde estuary in summer.

In the early 1960s I occasionally canoed down the River Spey to monitor breeding birds, and when I saw a goosander or mallard with young ahead of me I very carefully kept close to the other bank in order not to split up the brood. In the late 1970s we walked parts of the river to look for goldeneye broods, and the first complete survey was made on 17 and 18 July 1981, to estimate how many young had survived. We located 45 unfledged and 52 recently fledged young goldeneyes in 13 groups (one flock containing 22 birds), giving a total of 97 juveniles. It was unlikely that we found all the family parties; therefore 97 surviving young was a very good survival rate from a hatched nest total of approximately 286 young. A similar survey in 1982 located 76 young in July, and in late July 1984 at least 101 ducklings were located on the main river. On 24 July 1987 I walked from Aviemore to Boat of Garten along the River Spey to carry out a goldeneye survey and to look for suitable natural nest holes. Between 9.30 am and 1.30 pm, I counted 5 different females with broods of 1, 3, 3, 5 and 7 young, and 6 broods of larger young – 1, 1, 2, 2, 4 and 6 young. I also saw two female red-breasted mergansers, one with 4 young. I also noted two ospreys, 4 sand martin colonies, 350 holes in total, seven territories of common sandpipers, two of grey wagtails, one dipper, one red-throated diver calling overhead, a moorhen and 14 mallard. That day, I saw just six salmon fishermen, an indication how of quiet the river was in those days. In the 1980s there was a row between the traditional fishermen and the sports interests, and the latter won. In consequence, the river is no longer a quiet area for nature, but is instead a boisterous highway for canoes and rafts.

On 3 April 1984 I was invited by the Wildfowl Trust to help with an aerial survey by light aircraft of the River Spey from its mouth to Spey Dam. It was a superb, sunny day, and Keith, the pilot of the small Cessna, kept our flight path over the river with twists and turns. We counted 216 goldeneyes (additional birds would have been on various lochs), 51 goosanders, 17 red-breasted mergansers, 2 cormorants – and 69 salmon fishermen. The Wildfowl Trust carried out a similar survey on 12 February 2008, with Carl Mitchell and myself as observers. This time, we counted only 109 goldeneyes and 10 goosanders.

* * *

The first time we suspected that an egg collector had illegally removed a clutch of eggs was in 1976 at Loch Vaa. As we've seen, two clutches of eggs were taken in 1978 by egg collectors. From then on we noted that occasional clutches of eggs, and even the down, were stolen by humans. At this time the goldeneye was not on the specially protected list of breeding birds in the United Kingdom. In the 1970s I was a member of the Advisory Committee on Bird Protection for Scotland and subsequently a member of the Advisory Committee for the United Kingdom. Much of our debate in the late 1970s and '80s concerned the updating of the 1954 bird protection laws. I lobbied for the goldeneye to be added to Schedule 1 to give it special protection throughout the year. It was given special protection under the 1981 Act but only in the breeding season, leaving it as a legal quarry species in the hunting season. To combat egg thieves, the RSPB started a project whereby field staff started to mark rare bird eggs with ultraviolet pens to aid prosecutions. In 1986 Dave Pullan, one of my team's roving wardens, gained evidence of egg thieves taking goldeneye eggs in Strathspey. On 5 November 1986 Dave and I drove to Inverness courthouse for him to give evidence in the case against two egg collectors, although on the day they pleaded guilty and were fined £150 each, so Dave was not required in court. It was a successful conclusion to the case. We ended a good day in style, by checking out the Longman sewage outfall at Inverness and counting 300 goldeneyes.

As a side note on the legal protection of the species, I called to see an old friend, Stephen Frank, on 12 December 1991, to witness the official ringing of one of his falcons – I was then a part-time government wildlife inspector. Stephen was a famous falconer and an expert on peregrine falcons, which he bred and with which he hunted red grouse. He was a bachelor and his cottage on the moor near Dornoch was an interesting place to visit, with a scrum of English setters taking up most of the sofa. A hotpot was always bubbling on the wood-burning stove, and this time he invited me to have some lunch, with a quip that he had shot one of my 'protected ducks' when he was out hunting. The wings showed me that it was a young female goldeneye, and I remembered saying once that goldeneye would not be worth eating. Stephen produced two plates and some toast, put half the goldeneye duck and half a grouse on each plate, and I had to agree that it was tender red meat and tasted like pâté.

In 1984 Bob Swann found the first two pairs of goldeneye nesting outside Strathspey at Drumnadrochit, by Loch Ness, but he suspected that the first clutch of eggs was predated by pine martens. The next summer this was confirmed, when the eggs were again taken, this time definitely by a marten. In 1985 the first Strathspey nest was predated by pine martens. As the years went by, martens became regular nest robbers. This species had been mentioned to me, during my Swedish visit in 1979, as a major predator of goldeneye and goosander eggs. When I arrived in the Highlands in 1960, the pine marten was exceptionally rare, with the only regular small population in Wester Ross. I remember I failed to see the species between 1960 and 1963, despite searching in the western Highlands. When I returned to live in the Highlands, after seven years on Fair Isle, I became friends with Major Eric Hunter, who lived near Gairloch. He was a keen naturalist and was the first to encourage pine martens to come at night to his bird table by putting out raspberry jam and fresh hens' eggs. I remember staying with him and his wife in his cottage above the sea and, as dusk took hold, eagerly watching at his big window until, suddenly, there was my first marten, only a few feet from me, licking enthusiastically at raspberry jam.

Pine martens benefited in the early 1970s from a great increase in the planting of conifers with a subsequent explosion of field voles as well as a decline in the number of people killing martens as pests. Before long, they spread into Easter Ross and Inverness-shire. In the early 1980s they reached south of Inverness, then to the watershed of the River Spey, and from 1984 onwards pine marten predation of goldeneye eggs increased in our study area. Sometimes they even killed and ate incubating females; they also found the nest boxes ideal as dens and sometimes used them to rear young. We soon realised that the safest boxes were those on islands, in isolated trees and on buildings. There was also merit in regularly moving boxes, so that the martens were less likely to find them. I realised we were fortunate to have had the colonisation start at a time when this specialised predator of hole-nesting ducks was absent. In 1987 four nests containing clutches were taken over by red squirrels and the ducks deserted.

## Swedish goldeneye visit

In June 1979 I was granted a month's sabbatical by the RSPB, so I decided to visit Scandinavia and chose three study subjects: fish farming, forestry, and osprey and goldeneye conservation. This was also a very exciting chance to check out several species which had started to breed in Scotland, in areas where they were common. After visiting Sten Österlöf at the bird ringing office in Stockholm, I drove north up the Baltic coast with a contact name and address in my notebook. Many miles further on, I suddenly saw a road sign near Ljusne, so tracked inland and called at the local post office. My Swedish was non-existent, except for essentials like *knipa* for goldeneye and *fiskguise* for osprey. The lady told me she had just seen Owe Hedberg, the local ornithologist and bird ringer, walking down the road. I headed on and quickly caught up with him. He had no English but he got into my car and we spent the whole day birding, communicating mainly through the calls of birds I wished to see. He had about 50 nest boxes erected beside lakes and rivers in the surrounding forest land. The boxes were very variable in manufacture; some were just hollowed-out logs supplied free by the local forestry company. His nest boxes had been up for at least 15 years. In 1979, twelve boxes were occupied by goldeneye and four by goosander (a new colonist in his area and increasing). One nest box was occupied by a female goosander that had taken over a part-clutch of goldeneye eggs; both clutches of eggs had hatched.

The boxes were generally 4.5 metres up in isolated trees, and near each group a rough ladder was hidden in the bushes so that the boxes could be examined. The boxes were up to a kilometre away from the water in open forest; some were used by tawny and Tengmalm's owls. Owe's real interest was in ringing. He had found that the females could be caught easily in the nest boxes without desertion, and the ringing showed that females came back regularly to the same boxes. The ducklings remained in the nest for about 36 hours; he let all the ducklings hatch and dry out before marking them. In previous years he had marked them with standard duck wing-tags which he had subsequently identified on females when they started to breed. In 1979 he had been commissioned by Sten, at the ringing centre, to try a new method of duckling ringing pioneered by Russian bird ringers. A normal goldeneye ring was used for each duckling; a strip of adhesive foam strip (as used for draught excluders around windows) was cut and fixed inside the ring, before the ring was fixed on the leg, the foam being strong enough to stop the ring slipping down over the foot but soft enough not to interfere with the leg's growth. When the leg was fully grown the foam strip slowly sloughed off. He ringed a brood of goldeneye and a brood of goosander that afternoon, and I tried the novel technique, impressed by this unique method of ringing ducklings. I thought how marvellous it was to learn something new in such beautiful surroundings from a fellow birder, despite not being able to speak each other's language.

An area I was keen to visit was Värmland, and it had been arranged that I would stay for a few days with Erik and Carolina Jonsson at their farm in the forests north of Storfors. On the first evening, talking birds, I mentioned that I had not seen scarlet rosefinch breeding. 'They nest near our house – you'll hear them first thing!' they said. They told me I would hear the bird's distinctive call – 'pleased to see you ... pleased to see you' – and I certainly did. The local male woke me at 4.30 am and, once I'd heard it, I couldn't get that call out of my head. Their farm was surrounded by nature. There were ospreys, hobbies, beavers and elk – a paradise for a Scottish birder. It was also a study area for goldeneye, on forest land managed for game and wildlife by the Jägareförbundet Värmland (Swedish Sportsmen's Association). Sven Fredga (National Director for Waterfowl Management) had studied goldeneye there and run a nest box scheme for 30 years, since 1949. There were about 100 nest boxes in an area, about 5 square kilometres, of forest and small farms dotted with lakes, tarns, bogs and rivers. He had ringed females for many years and obtained much data on return and breeding success of individual females and the success of different boxes, all of which were code numbered. Since 1970 over a thousand ducklings had been wing-tagged with Swedish museum tags. In 1979 Hilary Dow had the exciting opportunity to carry out her PhD research on this population and its massive database. Sven and Hilary showed me their work and their field site, and I learned so much about their research on the breeding biology and nest box management for goldeneye.

The ducks return to the Storfors area, maybe some from wintering in Scotland, as the ice starts to break up at the beginning of April. The first eggs are laid in mid-April. The boxes were scattered in all sorts of localities but principally near the water's edge, about 2 metres above ground or water level. More use was made of tiny ponds on forest bogs than by the birds in Scotland. After the females have completed their clutches and been incubating for a week, it is safe to catch them on the nest, either in the box or by using a small hand net as they fly out of the box. Most of the ducks are now either ringed or tagged from previous years. A small number nest in natural holes, but the area had been pretty intensively felled for timber and old trees were scarce. In 1979 Hilary Dow found that 35 out of 100 boxes were used, which was fewer than usual – the very severe winter of 1978–9 may have killed some of the breeding population. Twenty ducks were successful and a total of 161 young had been tagged (none had been missed). Seventeen was the biggest number of young from one box, from probably more than one female. Six clutches were abandoned for unknown reasons, pine martens predated five, and other goldeneye ducks interfered with four boxes. The drakes leave at the beginning of June and, when I was present, young were leaving their boxes regularly. The date of hatching can be gauged by checking the eggs for the first signs of starring. Young are tagged at one day old; all are removed from the box and placed in a special container full of goldeneye down. They are tagged and then returned to the box. Care must be taken to let them quieten down so they do not explode, scattering in all directions.

Sven Fredga told me that 50 to 100,000 boxes had been put up for goldeneye in Sweden; many had been made by forestry and hunting organisations, as well as by birdwatchers and local people. One firm had sold 10,000 boxes in recent years. Not surprisingly, the ducks' numbers have increased considerably in the last two decades.

I was very impressed by the goldeneye studies I saw in Sweden. The habitat was very similar to the Scottish Highlands, and there was no reason why the species should not continue to increase in Scotland. My visit encouraged me to increase our nest box project and gave me new ideas for the design of nest boxes, their siting and density.

I saw goldeneye in many areas of Sweden: the mosaic of lakes and rivers surrounded by pine and spruce forests is ideal for this species. Females were seen on fresh water from the largest lakes to the smallest ponds; the females give a distinctive call when they are off their nest, and I saw and heard them frequently. In Denmark, I saw 17 drakes on a lake in Rold Skov forest, north-east Jutland, on 11 June; presumably pre-moulting drakes from Sweden, but a small Danish breeding population had become established. I saw several nesting goldeneye north and east of Oslo in Norway, but nest boxes were much less common there.

There was a fascinating spin-off from this trip when I talked with foresters about my findings. I remember their disbelief when I showed them pictures of some of the areas which had been felled by their Swedish counterparts. The common practice there at that time was to leave any dead standing trees in situ, rather than cut them down, and to put up tall T-perches. When I asked a Swedish forester why, he told me that it was really good for providing perching places for birds of prey and owls, so that they could keep down the numbers of rodents that eat seedling trees. I was very taken by the idea of maintaining standing dead timber and recommended it for use in Scotland, but my forester friends were not in favour: most of them preferred to be tidy. It was many decades before one of those young foresters was in a position to give it a try, and it's now standard practice in Scotland to leave dead standing trees in clear-fells. It's great for wildlife conservation.

## Scottish peregrines for German recovery projects

I was working at the RSPB Operation Osprey camp in Strathspey in 1960 when, on 18 June, George Waterston arrived after a visit to Orkney. He handed me a cardboard box, and when I looked inside, there was a young peregrine falcon, staring back up at me and hissing. Confiscated by the RSPB from a person in Orkney who'd been holding it illegally, it now needed someone to rear and release it. I constructed a wooden hutch for the bird and fed it on fresh meat. I trained it to perch on my hand and to fly to me to feed, like a falconer's bird. I called it Desmond. On 25 June it flew off over the forest and I thought it had gone, but it returned – very hungry – three days later. I ringed it on 1 July and, soon after, released it on the nearby farm fields. Desmond circled above me and, all of a sudden, was gone, flying strongly over the forest. I never saw it again and, sadly, George later heard that it was found on 31 July at Birnie, near Elgin, some 45 kilometres northeast of where it had left me. Its wing was broken and it had probably been shot.

In 1962 another young peregrine was sent to me for rearing and release. That one I called Rannoch, from its Perthshire origins. Again I found it fascinating to watch a falcon grow, to see it learn new ways of landing on my wrist and observe its individual character. I was able to release it to the wild from the same field and it flew away to the west, without a backward glance.

Juvenile peregrine falcon being reared in June 1960 for release at RSPB Operation Osprey camp near Loch Garten.

Peregrine falcons are one of the most widely distributed bird species in the world. They live throughout Europe, Asia and North America, reaching as far south as the Falkland Islands, as well as being present in Africa and Australia. The northern populations migrate to southern regions for the winter. They are extremely fast-flying birds and their normal prey are birds caught and killed in fast flight. They do not build their own nests but nest in cliff faces, hollow trees, the tree nests of other bird species, on the ground and – nowadays – on structures such as cathedrals, large towers and bridges. Peregrines have been used by man for falconry for over 2,000 years. In some parts of the world they have been subject to human persecution; there was a special case in the UK during World War II, when peregrines were killed along the coasts, especially in southern England, so that carrier pigeons could return safely with news of wartime operations.

I always enjoyed watching peregrines in the wild, so in the early 1960s I was stunned to observe an unfolding tragedy in the UK, mainland Europe and North America. The new chemical pesticides created for agriculture in the 1940s and '50s were having a dramatic impact on the falcons' breeding performance, due to concentrations built up in their bodies. The peregrines were feeding on smaller birds which had eaten insects and grain contaminated with persistent and dangerous chemicals, which became concentrated in the peregrines further up the food chain. Peregrine eggs became thinner-shelled and infertile, and even adult birds were dying of pesticide contamination. The population started to plummet, particularly in England and southern Scotland, with only the population in the north of Scotland holding its own, although even those birds were being affected. I remember visiting a peregrine eyrie just north of Carrbridge on Dava Moor in the spring of 1961. It's a beautiful, remote site, which looks out over miles and miles of heather moors. The nest scrape held three eggs, but I noticed that one was leaking and had a small crack in it. When I went back a few weeks later, all I found was a failed nest, so I removed what I could for scientific analysis by the Nature Conservancy.

It was at this time that Dr Derek Ratcliffe, the peregrine expert, was seconded from the Nature Conservancy to the BTO to organise and carry out the first UK-wide census. In May Derek came and spent a night at the osprey camp at Loch Garten, and the next day I joined him on a field trip, looking at nest sites in Badenoch and Strathspey. Later, his survey report revealed the true extent of the peregrine's decline, and the link to organophosphate pesticides was established. Finally, some years later, the most dangerous chemicals were banned and birds of prey started to recover.

In North America the extent of the damage was even greater and the species disappeared from many areas of the United States. Tom Cade of Cornell University headed a recovery project, using young birds reared in captivity. The first person to be successful in rearing peregrines in captivity and releasing the young was Dr Christian Saar, a veterinary professor living in Berlin at the time of the Cold War. He had started a breeding programme in Berlin using peregrines from several sources, and released the first eight young from a big building at Tempelhof airport in 1977. The previous year, he was speaking at an international conference on peregrines in Abu Dhabi and mentioned the difficulty of obtaining nominate peregrines, European stock, for breeding. Professor Vero Wynne-Edwards was in the audience and suggested he try Scotland, where he was chairman of the Advisory Committee on the Protection of Birds in Scotland (ACPBS). Dr Saar applied for a licence for two males and two females. I was also a member of the ACPBS at that time, and his application came before our committee in February 1978. My diary says that I had lunch with Dr Leo Batten, the bird-licensing officer based at the English headquarters of the Nature Conservancy, and that at our meeting Wynne-Edwards and the committee supported the project. Our decision that day led to the award of a licence.

The four young were collected by Dick Balharry of the Nature Conservancy in the summer but, unfortunately, all were females. Nevertheless, they added to his breeding project, which started to show success and the birds began to breed in East Germany, which at that time was not accessible to Western ornithologists. Over time, this population thrived and returned to the traditional nest sites in riverine cliffs and mountains.

In the pre-pesticide era, the East German and Polish Plain, which was covered by extensive forests and had no cliffs suitable for peregrines, held a breeding population of possibly up to 1,500 pairs, which used the old nests of eagles, ospreys and kites rather than cliffs. This tree-nesting population had been completely lost through deaths and breeding failures in the pesticide era, and the culture of nesting in trees was lost. Christian Saar wanted to change this. I met him several times in the Scottish Highlands, when he came to hunt red grouse with trained falcons, along with falconry friends in the UK, whom I knew.

On one of those days, in the autumn of 1994, Christian asked me about the chances of obtaining a few more peregrines from Scotland for the breeding programme in Germany. His 1978 breeding females were very old – two females had become expert breeders, producing 55 and 47 young for the peregrine recovery programme, which had reached more than 50 pairs in the wild. The oldest finally died in 1995 at 16 years. I was very keen to help, but this was quite a difficult time for conservation relations between Scotland and Germany, because we had had serious cases in 1993 and 1994 of German nationals stealing peregrine falcon eggs from wild nests in Scotland for illegal falconry. One case near Inverness had involved me appearing in court in Mannheim to give expert evidence in a criminal case against two illegal falconers. Nevertheless, I knew that this was an important project and I applied, on Dr Saar's behalf, to the Nature Conservancy in Edinburgh for permission to collect three pairs of young peregrines for the German project. The licensing officer, Tony Colling, was supportive, and Elizabeth Dudgeon of the Scottish Home and Health Department issued me a licence in the spring of 1995. Part of the reasoning behind the approval was the fact that the previous young peregrines donated by Scotland in 1978 had produced, in captivity, such a remarkable number of young to help with the re-establishment of peregrines in northern Germany. These new birds were to help with the exciting new project to restore tree-nesting peregrines.

In spring 1995, when we were carrying out our monitoring and protection of peregrines, I looked for suitable donor sites and asked landowners for permission. At that time there were 130 pairs in the whole of the Highlands and nearby islands. During the national census in 1981, which I organised in the north of Scotland, we checked 271 traditional nesting sites, some of which were long abandoned, and found 144 pairs. Ten years later, in 1991, we located 129 pairs. Sadly, since then the situation has further declined, in contrast with southern Scotland, England and Wales, because of the decline in wild prey in the north and west Highlands, as well as increased levels of illegal persecution on grouse moor estates.

On Saturday 17 June 1995 I set out to collect the first peregrines from a cliff eyrie not far from my home near Nethybridge in Strathspey. In the early afternoon, I drove by Land Rover along a farm track, with an old farmer friend, Jock Rattray. Leaving the vehicle at a deserted house, we walked out from the birch wood across a heather moor, passing noisy common sandpipers nesting by a small loch on our way. It was a nice walk up through the lower hill scattered with occasional Scots pines, seven red deer running off ahead of us, and then we scrambled up steeper ground and screes to reach the nesting crags. Both adult peregrines were present and calling noisily at us, the female at this nest being a very large individual.

I climbed up to the eyrie, which that year was in the western ledges by a small rowan tree – the cliff is only 5 metres high and about 20 metres long, so there was no need for a rope. There were three big young in the nest and two of them flapped off one end of the ledge, landing in long heather close by. I caught them both and put them in two small rucksacks; they were both large, well-fed females, and had recently eaten a dead rook. As we walked downhill, two ospreys were fishing on the small loch, one of them catching a trout and flying off east. I took the falcons home and kept them in a special compartment in a barn at the farm.

Five days later, on 22 June, I collected another two young peregrines. It was a hot, very sunny day, perfect for fieldwork. In the early morning I went to Moray, about 60 kilometres north of my home, to monitor osprey nests and later collected a friend, Lorcan O'Toole, at a pre-arranged meeting place at the local Little Chef car park near Forres. Lorcan, from Donegal, who at that time was working on our red kite reintroduction, has since returned there, and led the successful project to restore golden eagles to Ireland. We called at the house of the estate keeper, Jimmy Maclean, and he came along with us, leaving us above the riverside nesting cliff while he went on with his vehicle and walked across the River Findhorn, so that he could help with directions from the other side of the river. The female was calling above us as we climbed down the bulk of the cliff through scattered trees, stopping at the top of the sandstone outcrop.

Female peregrine and her brood of young at a monitored nest site in Inverness-shire, June 1975. This nest was featured in the BBC film *The Shadow of the Falcon.*

Fixing ropes, Lorcan climbed down one side of the sandstone cliff, while I went down the other. The young peregrines started running along the sandstone ledges from their hidden eyrie. Lorcan ringed the female chick and then caught the other two, which were tiercels (males), and passed them up to me. We climbed back up the cliff, where the keeper collected us in his vehicle to go back to his house for tea. He told us that two people had turned up on his side of the river and asked what was happening. Jimmy said that we were ringing kestrels. 'They're not kestrels, they're peregrines!' they retorted. Jimmy agreed and tried another tack, saying that it was Roy Dennis of the RSPB ringing peregrines. 'He doesn't work for them any more!' came the reply. Jimmy tried a third tack, and told them to leave.

Unfortunately none of the other nests for which I had permissions had young, so I looked after the four birds at the farm until they were collected for transit to Germany. Two students from there, Katrin Bednarek and Michael Greving, arrived to take the young peregrines back to Hamburg, where Christian Saar had his peregrine breeding facility. As they set off, I gave them letters of authority and telephoned the customs officers at the ferry port in England, to make certain there were no problems with export. They were all marked with special Department of the Environment cable-tie rings to prove their identity.

The remaining licence for two peregrines was activated by the Nature Conservancy for 1996, and I decided to take the young from a different area. On the afternoon of 15 June I drove north to Forsinard in Sutherland to meet David Clements, the local gamekeeper, who knew a suitable nest. This area of rolling moors, about 200 kilometres north of my home, is called the Flow Country, a wide expanse of heather, peat and mosses with many lochans, superb for wading birds such as golden plover, greenshank and curlew. We drove off in his vehicle and then walked over open moorland to a peregrine eyrie in a rocky gully, a very remote location. There were three rather small young in the nest and I took the two largest ones, hoping for a pair, although both later proved to be males. The prey remains included red grouse, blackbird, starling and feral pigeon. It was a beautiful sunny evening as we walked back, and we could see for miles across the moorlands, with curlews, snipe and red grouse calling and, in the distance, a greenshank giving its distinctive flight calls. On the drive back south, I stopped to look at several osprey nests and finally got home half an hour before midnight. On the following day, I walked up to the nest near my home, from where I had collected two young in 1995. The pair had been successful again and there were three large young in the nest. Both of the adults were away hunting, so I could ring them without the accompanying noise of angry parents. Prey remains were cuckoo, lapwing, song thrush, feral pigeon and black-headed gull. The two young from Sutherland were kept in Scotland until 11 October, when Dr Saar transported them to his breeding station at Hamburg.

Later in the summer, Christian invited me to visit him in Germany, so on 9 August 1996 I caught a plane to Hamburg. He collected me and took me to his house in the leafy suburbs before visiting his peregrine breeding station, which I found very impressive. I had a look at the two pairs which I had collected in Scotland the previous year – they were looking really well in adult plumage, and one of the males was very white-fronted. Christian showed me a peregrine release cage on the very top of a big Douglas fir and also the pulley system that took fresh food up and by means of a neat little basket tipped the food into the nest for the chicks. We had a short walk in the forest to look at a goshawk's nest, where two flying young were calling through the trees.

The next day we drove east to Himmelspfort, crossing the old border between East and West Germany, and then had a long drive along forest tracks to Woblitz field station, passing two active osprey nests on pylons. There we met Paul Sommer, who ran the place, and my osprey friend Daniel Schmidt. After refreshments, we crossed the small river by boat – Paul's tame crane coming on the walk with us – and I was shown a hobby's nest in a basket in a dead tree. The two adult and four young hobbies were flying around catching dragonflies. We then made a short walk to a peregrine release site, an old, man-made osprey nest at the top of a huge Scots pine. This was the hack site from where two young had recently fledged, fitted with video camera gear – unusual in those days – to maintain a record of the event. We hiked up onto a big sand dune in the forest to a high seat and could hear the two young peregrines calling from near the release site. Then the adults arrived, the male bombed the young female and the adult female took food from the platform.

Next they had a special event for me further through the forest, where I was shown the site where the very first pair of tree-nesting peregrines had bred in an old black kite's nest in a Scots pine beside a small lake. Both adults were two years old, colour-ringed birds released in 1994, and the team were very excited that their tree-nesting re-establishment project was really underway. Christian had reported the news to me in May, his letter revealing how it felt to have succeeded: 'This is the world premiere of the beginning of a new peregrine tree-nesting population after the crash 30 years ago.' From our viewpoint we also saw two red kites, black kite, sparrowhawk and honey buzzard. Next we visited the other hack site, where three young had been released; the CCTV at the nest was linked by cable to a monitor in the field station. We stayed there that night after a barbecue of turkey.

On 11 August we drove further east towards the Polish border, taking the motorway south to Dresden and then heading across country to Radeburg and finally to Rackenwalde. There we met Hans Ebert and his wife in their lovely wooden house overlooking the Sächsiche Schweiz, a landscape with amazing sandstone rock buttresses carved by the River Elbe. Ulrich Augst from the National Park joined us, and we set off after tea to the river and out to an isolated rock outcrop, where a pair of peregrines had reared three young that summer. The adult male was perched near the used nest, which was in a natural hole in the rock face. These peregrines were special to Christian because, when he reared and released the first birds in West Berlin many years before, these were the locations where they chose to breed. Finally, after the fall of the Berlin Wall, he was able to visit all these places and meet his friends, the local peregrine enthusiasts. Next we drove into the forest to look down at an attractive site where young had been released from a hack site in the cliff, with the food being dropped down a plastic pipe onto the nest. An adult was present but peregrines had not yet nested there – this bird often caught swifts. We stayed the night with the Eberts and had a lovely evening of talking about raptors, conservation and the wildlife in this beautiful part of the country. We ate a variety of delicious meats and sausages with homemade beer, and a lovely evening was finished off with schnapps, well after midnight.

The next morning we were away early to meet Ulrich. He drove us up through a beautiful forest of pines, spruces and beech trees to a plateau close to the Czech border, and out we walked through blaeberries to a cliff edge set in dramatic scenery. We sat on the point, over grey sandstone buttresses with gnarled trees growing on the ledges; a pair of peregrines and one of two young were flying about, screeching at our presence, while crested tits called in the trees above. Ulrich told me that there were about 12 pairs of eagle owls in his area, as well as pygmy and Tengmalm's owls. Before setting off home, we visited another peregrine nest that had had two young, and I was told that seven young had been reared from three pairs in Sachsen that summer.

In October 1997 Professor Wolfgang Kirmse invited me to give a talk about 'Scottish peregrines and their contribution to recovery projects' to a German peregrine conference at Prerow on the Baltic coast of northern Germany. While it was very nice to meet old friends and talk peregrines, it was particularly special to meet again Werner and Heide Eichstädt, whom I had originally met in 1980 in Mongolia. They had invited me to stay with them in northeast Brandenburg, close to the Polish border, in 1984, at a time when East Germany was politically very challenging for a Western birder. Despite such difficult times, we – as ornithologists – enjoyed four superb days of birding.

Over the years I kept in touch with Christian, especially when he came to Caithness for his annual hawking. In 2000 his Scottish

Peregrine chicks hatched at Christian Saar's Peregrine Project near Hamburg for release in Poland, May 2018.

peregrines produced 10 young, all released in the tree-nesting area; in 2003 they found six pairs nesting in the wild in trees. In 2004 there were 10 pairs of tree-nesters and the Scottish birds produced 14 young for release into the wild, either by tree hacking or fostering. In 2006 they identified a male breeding in that area whose grandfather was one of the Scottish-donated peregrines. In 2008 the two Scottish pairs were still producing fertile eggs at 13 years old, and seven young were released.

In late May 2010 I was invited to a celebration of the success of the tree-nesting peregrine project, where we met Christian and our osprey friends, Daniel Schmidt, Paul Sommer, Tadeusz Mizera and Torsten Langgemach. It was a very interesting and congenial gathering, listening to the success story that was the restoration of the peregrine to the forests of the great plain of northern Germany. On the field trip day, we watched four young

Christian Saar and the author in Hamburg, May 2018.

peregrines in an old osprey nest basket in a big Scots pine, the adults very noisy. We visited various osprey nests in that area and – a real pleasure for me – saw a male ortolan bunting singing from a single tree beside a field of rye.

It was another marvellous visit to Brandenburg. It made me proud to think that Scottish peregrines were helping to restore the species in Germany and that I had been able to help. I was pleased that my fieldwork in the Scottish Highlands had helped this ground-breaking project, and that I had made some very special nature conservation friends in Germany. In May 2018 my family and I enjoyed a lovely weekend break in Hamburg with Christian and his wife Christiane, and learned that the new peregrine chicks from his breeding centre were destined for release in Polish forests, to ensure the spread of the tree nesters. This year the 1,000th young peregrine was released, a young male from Scottish parents. Christian's lifetime contribution to the future of peregrines has been immense. He is a very special falconer, ornithologist and friend.

Red kite. (*Photo by Mike Crutch/A9Birds*)

# 3

# CHARISMA

## *Red kites flying free*

Sometime in 1953 a small group of Boy Scouts from Park Gate in Hampshire headed to mid-Wales in their Scoutmaster's Austin 7 car. We were on an expedition to recce the location of that summer's camp and hoped, en route, to see a red kite, unknown at that time in England. We headed for Tregaron and across a very remote moorland road, the mist hanging low, towards Devil's Bridge. Somewhere on those bleak moors we did glimpse the near-mystical bird, far in the distance and, looking back, it's hard not to see it as a talisman, or a sign of what my future would hold. Later that summer we did a 15-kilometre hike each way from our scout camp at Pantglas Farm, inland of Machynlleth, to the summit of Plynlimon. That day, we weren't so lucky – we saw no kites – and I now know that the location of the 14 pairs in mid-Wales at that time was kept top secret, especially from 13-year-old boys. Back then, though, I was happy with my first wild polecat – another very rare Welsh speciality of the day – and it would be nearly 20 more years before red kites played a part in my life.

On 9 January 1972 one arrived unexpectedly at Gaskbeg, north of Dalwhinnie in the Scottish Highlands. My friend Doug

Weir had good views on 13 January, and so, next day, I went over early to meet him. Doug put out his tame buzzard as a decoy beside rabbit carrion at Balgowan. We sat overlooking the location for an hour, but there was no sign of the bird, so we drove to Gaskbeg Farm for a chat and coffee with the farmer. Doug's wife, Penny, arrived with news of the kite on the other side of the river, but when we got there it had gone. I was back on 15 January and, after a lengthy search, Doug and I saw the red kite soaring over a hill, giving us the briefest of views; fortunately, an hour later, it reappeared with a buzzard over Cluny Wood. I wrote of its plumage and behaviour in my field notebook, concluding with the comment: 'It was a beautiful flyer.' It wintered there until at least 5 February 1972 and was most likely a vagrant from mainland Europe. In Wales that year there were 29 pairs.

Whenever I dipped into the old bird books I saw that the red kite used to be a common bird in Scotland. Just 45 kilometres downriver from my 1972 sighting, the Reverend Forsyth, the minister at Abernethy Church in Strathspey, wrote in 1885, 'I have not seen a kite for many years. I remember them when they were very numerous and a pretty sight it was to see them hunting for mice in the stumbles on a fine autumn day, their mode of flight and of striking prey was very beautiful.' In Scotland they were called the gled or salmon-tailed gled. Baxter and Rintoul, writing in *The Birds of Scotland*, recall: 'The second half of the 19th century saw the virtual extirpation of the kite in Scotland. The fact that they were carrion feeders makes them very easy to trap or poison and there are all too many records of the parent birds also being shot at the nest and the eggs and young taken. Kites were ready to approach human dwellings and would thus render themselves vulnerable. There are many reports of them coming to the farms and houses in the country districts and taking the young of chickens or stealing clothes drying on the grass or pieces of paper for their nests.' The old vermin lists show the ruthless war waged on them; in the Callendar hills between December 1824 and December 1825, 105 kites were killed; while in Glengarry, 275 were slaughtered between 1837 and 1840.

Most of the last breeding attempts in the 1870s and '80s in Scotland were documented by Harvie-Brown. In 1878, Edward Booth, the Victorian egg collector from Brighton, found kites breeding in the Rothiemurchus district and said that he knew of them nesting in the same locality at least a year or two later. The last proven nest was at Choire Dhomdain, Kintail. In 1917 Mr Matheson said that a pair appeared and nested in one of the old kite trees at Glengarry. Professor James Ritchie wrote, 'It is almost impossible to believe that a bird once so common that its vast numbers on the streets of London excited the wonder of foreign visitors in the reign of Henry VIII should have suffered so grievously that in 1905 the few survivors in the British Isles could be counted on the fingers of one hand.'

We had lost the red kite as a breeding bird in Scotland and England, and when I lived in Inverness-shire in the 1960s there seemed no prospect of it returning, although occasional migrants from mainland Europe were recorded. Then, however, Don Smith, a keen birder in Ayrshire, recorded a colour-ringed juvenile red kite from Wales found shot in Kirkcudbrightshire in autumn 1975, just months after fledging. I remember talking with Don about this intriguing record at the annual birding conference of the Scottish Ornithologists' Club and wondering if and when we could get them back.

At this time, the sea eagle reintroduction was underway and I was determined that we should run a similar project for the red kite. It was a very rare species in the British Isles, and the Welsh population, although increasing, was subject to egg robberies and other misdeeds. More importantly, Peter Davis, the NCC field ornithologist and my old boss at Fair Isle, proved that the kites were living in a land of impoverished food availability and had been for decades. When he fed white rat carcasses to a pair, their breeding success immediately increased. Some time at the beginning of the 1980s, while at one of our regional officer conferences at the RSPB headquarters, I discussed red kites with my friend Roger Lovegrove, the Regional Officer for Wales. I knew that there was talk about taking the first clutches for artificial incubation and to rear chicks, in order to foil egg thieves and to encourage pairs to lay second clutches of eggs. By that time there were over 40 pairs, so I asked if it would be possible to move some incubated eggs to the Scottish Highlands. I did not get far, as Roger laughingly said, 'No way! It's our best bird in Wales and you already have enough rare and exciting birds breeding in the Highlands!'

I never stopped dreaming and scheming, though, that the red kite would follow the sea eagle and return to Scotland. In September 1983 I was birding at Falsterbo in southwest Sweden, an amazing place in autumn for observing bird migration en masse. I knew the Bird Observatory wardens, Lennart Karlsson and Karin Persson, through my Fair Isle Bird Observatory Trust contacts, and later helped them with the translation of an English version of their observatory book. Their gift to me for that help was a lovely painting of a migratory robin at Falsterbo bird observatory, which hangs to this day above my fireplace. As well as birds, there were the biggest numbers of birders I had ever seen in the field, and some were exceptional raptor experts. There I met Gunnar Roos, Nils Kjellén and Johnny Karlsson, a lecturer in the Ecology Department of Lund, who introduced me to Magnus Sylvén, who later became a good friend. I learned that he was carrying out PhD research on the red kites of Skåne and, as we discussed his studies, I mentioned my wish to restore them to Scotland. I was amazed to hear him say, 'Well, you can always have some from my study population.' 'Aren't they important for your scientific studies?' I asked. 'What you are aiming to do is more important!' he replied. What a special man he was. Projects – and lasting friendships – often depend on chance meetings like this.

After my Swedish birding trip, I decided I would start a project to restore red kites to Scotland, but knew it was not going to be easy. My RSPB colleagues were, at that time, very focused on the Welsh red kites, which were using up a lot of time and money in RSPB species protection, combating egg thefts and persecution, as well as carrying out research and working on improving breeding success. I started to talk to colleagues about the possibilities, but did not receive much support; in fact, several were completely against the idea. On one occasion I was told by a senior colleague that my request to release red kites might encourage cage-bird enthusiasts to breed and release black redstarts in London. There was no doubt that reintroductions could polarise opinion among close friends.

Outside the RSPB it was much easier. My farmer neighbours on the Black Isle, Bobby MacDuff-Duncan and his wife Kitty, a keen birdwatcher, immediately said that an area I thought ideal for release cages at the wooded end of their farm would always be available, if I got permission to start. Meanwhile, I occasionally met up with Flight Lieutenant Steve Rooke, an RAF Nimrod Captain at RAF Kinloss, whom I had met on the white-tailed eagle flights from Norway to Scotland. He promised that if I needed young red kites flown from Sweden to Scotland, he would try his best to build the trip into future training flights. I also had promises of plenty of rabbits for feeding the young, and Colin Crooke, my RSPB colleague in the Highlands, was totally supportive. There was no doubt in my mind that the Black Isle, where we had our RSPB office, was an ideal location: a rich mix of farms and woods on low, productive farmland, lots of farmers and landowners whom we knew, a high common buzzard population with above-average breeding success, and loads of rabbits. I was determined not to take no for an answer.

When the sea eagle reintroduction started to show success in 1985, there was a slight shift towards my idea to restore red kites to Scotland. Unfortunately, there was opposition to using the Black Isle, close to my home and office – my boss in Edinburgh thought we would spend even more of our time out of the office! Some colleagues argued that the kites should be returned to the locations in remote glens where the last pairs had nested before extinction. I and others, including the raptor expert Dr Ian Newton, argued that the last places were not necessarily the best habitat for a species but were rather the places where the birds had survived simply because of remoteness. The thrust of persecution usually starts in the strongholds, where a species is common, and only has the final impact on the outlying survivors at the very end. This is a truth throughout the world, for large raptors as well as for mammal carnivores like lynx and wolf, which all find themselves pushed into the least productive areas. A review of regional nesting densities and breeding success of common buzzard confirmed that the Black Isle was as good as anywhere.

Another two years went by until, in 1987, the end was finally in sight and – most importantly – we had decided, during prolonged discussions, that instead of a single release in Scotland there should be another in southern England, followed by a rolling succession of future releases to join up the new populations of red kites. It was an important decision. It was to be a joint RSPB and Nature Conservancy Council project, and our first meeting was on 20 October 1987, when we met at the NCC offices in Peterborough. There was Richard Porter and myself from the RSPB, Leo Batten and Mike Pienkowski from the NCC, with Peter Davis and Peter Walters Davis, the NCC kite experts from Wales, and Ian Newton from the Institute of Terrestrial Ecology. My diary records that it was a very good meeting and a go-ahead was ratified.

Our next meeting, in February 1988, started to put flesh on the bones of the project and we decided that the southern release should take place near London, while the northern would be on the Black Isle near Inverness. On 6 October I caught the London flight from Inverness to Heathrow, where I was collected by Richard Porter and Graham Elliott of the RSPB species protection unit. We met the rest of the Red Kite Project Team at Windsor Great Park and had a very interesting day viewing the area, much enhanced by Ted Green, who told us the histories and ecology of many of the ancient oak trees. There was no doubt that, ecologically, it was a good release site, but we soon learned that the head keeper, the factor and others had reservations. Fortunately, the southern members of the group worked hard and located a very suitable alternative site in the Chilterns, owned by the Getty family. On 24 January 1989 the group met in Inverness and worked through issues: 'some good, and some problems', as my diary tells me. The next day I showed the group the proposed release site, foraging areas and future breeding areas on the Black Isle.

Once the budgets were approved, we started getting ready for the year ahead. Between 1968 and 1988, red kites had occurred on migration in Scotland in 14 of the 21 years, just one to four individuals per year, except for 1988 when at least 10 were recorded An important task for me was to arrange all the official permits and licences, which I started to do in March 1989. On 3 April I applied to the Department of the Environment in Bristol for a CITES import permit, which is the international regulation to control trade in endangered species. Our project was for nature conservation, not trade, but we still required the CITES permit. Under the Animal Health Act, I applied to the Animal Welfare Branch of the Department for Agriculture and Forestry in Scotland for a licence to import red kites and to arrange with the government veterinarian in Inverness to inspect our premises for quarantine of the kites before they arrived, as well as the kites themselves before release.

RSPB Headquarters were to oversee the budget and considered that we, in the Highlands, were lacking in knowledge of red kites. This came as news to us, but we were happy for Dee Doody to come up from Wales to carry out the day-to-day work involved in the first year. He had been working with red kites for many years in Wales, was a first-class fieldworker and a talented artist. We had many things to do before the young kites would arrive in June, so Dee needed to familiarise himself with the Black Isle. Not long after his arrival, he called into my office and told me he was worried that the project wasn't going to be successful. I asked him why, and he replied, 'I've covered a lot of ground on the Black Isle and I cannot find suitable breeding habitat. Our research in Wales shows red kites really need mature oak woods on the sides of valleys; in fact our analytical models suggest they need mature oaks of a special age, spaced so many metres apart, on a slope of so many degrees.'

I agreed it was very different to Wales but suggested that those kites were a special case due to isolation, because my experience of red kites breeding in mainland Europe, especially in southern Sweden, indicated to me that they had no such specialised needs. And that was how it turned out.

We constructed hacking cages, on a wooded slope on Bobby's farm at Teandore, across the Beauly Firth from Inverness, with a great view over the valley to the woods. I had worked out a new design, loosely based on the original Fair Isle sea eagle cages, but decided to have the back wall wood-boarded so that we could approach the kites without them seeing us. Three individual cages were arranged side by side, with feeding sleeves fitted in the back wall so that food could be dropped onto the nests inside. Peepholes allowed us to view the birds and the nests to check how much food had been eaten. Each cage would hold three young, and the dividing walls were also made of wooden board, so that each 'brood' grew up without seeing the other young kites. The kites could be released by lowering the front of the cage by a cord tied at the back. We were ready.

On 21 April I got the all clear from the RAF, after a good planning meeting with Steve Rooke and Dave Pocock at Kinloss RAF base. They were ready, too. In Sweden Magnus Sylvén wrote to say that his colleagues Nils Kjellén and Ulf Sandnes would meet me when I arrived in Lund. On 3 May, when it was clear that donor young for the southern release site were not forthcoming from Wales or France, I wrote to the Import of Birds Section and received a variation to allow four of the young red kites to be taken from RAF Kinloss to the Wormsley estate near Henley-on-Thames, for the southern location. On 2 June, along with the CITES export licence from the National Board of Agriculture in Sweden, I received special authorisation to import the kites to RAF Kinloss instead of the normal designated ports of entry. Armed with the essential paperwork, I was ready to travel.

On Friday 9 June I flew from Aberdeen airport on an SAS DC9 flight over the North Sea to Stavanger. After half an hour, the plane continued to Copenhagen, where I caught the bus to the ferry port. As the boat crossed the sound to Malmö in Sweden it was cold and rainy, but soon I was on the train to Lund, where I met Nils Kjellén. I stayed at his home and we yarned about the project and raptors until very late, Nils being a great expert on raptor migration at Falsterbo. The next morning, Saturday, it was a beautiful sunny day when I was collected by Ulf Sandnes and the red kite ringer Per-Olov Andersson.

First of all, they drove me to the Lund University Field Station at Stensoffa, which Johnny Karlsson, a lecturer at the University, had organised as my base. I found a room, dropped my gear and quickly we went to the field. The first red kite nest we visited was in an 8-metre-high birch tree, just 50 metres from the public road and close to the farmhouse. It was a real surprise: I was clearly on a very exciting mission. The young were too small, though: I was looking for young of a certain size, well feathered and capable of maintaining their body heat without being brooded, able to pick up their own food but not so big that they might fly too soon. It was important to keep siblings together and to grade the young by size so that each 'brood' reared in Scotland would be at the same stage of growth. We therefore moved on to the next nest, on the Revingehed army ranges. Per-Olov knew all the nest sites and I had been invited to join them on the annual ringing visits to their monitored nests. Ulf was a very competent tree climber and found three young in a big nest in a 20-metre Scots pine. The brood was lowered gently to the forest floor in a sack; my notebook of the day says, 'What brilliant little ginger-red creatures.' I chose the largest chick while Per-Olov ringed and measured the other two and sent them back up to Ulf to place in the nest. This was to be our routine for the next couple of days.

The next nest had two young, both too small to ring, but the fourth nest, in a 27-metre-high beech tree, had three big young, and I collected the smallest one. The next nest was in a similar high beech and I was impressed by the Swedes' climbing technique, involving a long aluminium jointed pole with steps that was hooked over a branch. It looked dodgy to me, but Ulf was a skilled climber. We again found three large young and I collected the smallest. We stopped for coffee and lunch on the roadside and listened to the local birds – redstarts, blackcaps, tree pipits and thrush nightingales – while hobby, kestrel and common buzzard flew overhead. After lunch, we visited three more kite nests, and one in a really big beech tree contained a superb brood of four young, of which I collected two. Before returning to the field station we stopped in Sjobo for me to buy groceries as well as fish and meat for the young kites. I used a lab room to house the kites and fed them on cut-up meat and fish before I prepared my own supper. Before dark I walked down through the woods and marshes to the Krankesjön Lake, where I watched marsh harriers and black terns and listened to marsh warblers and thrush nightingales. I felt I was in heaven as the songs of the nightingales rang through the summer night.

Before being collected by the team at 7 am the next morning, I got up very early to cut up food and feed the young kites. Nils had received a message about a young kite being found below its nest, so we headed to a house on the Baltic coast, only to find it was, in fact, a young common buzzard. We left it with the keeper and returned inland to Fyledalen, which was to become a favourite valley of mine, because there was such a lovely view across the woods and fields of Skåne. While we sat there a honey buzzard mobbed a red kite, while icterine warblers sang in the trees and marsh warblers called in the valley below. The second nest we visited was in another big beech tree by a spruce plantation, where Ulf lowered a terrific brood of four young and I collected the two middle-sized ones; Ulf reported that the nest was full of trash from a local dump. That day we visited seven other nests, some in trees too difficult to climb, some with small chicks, and two nests with three young, taking the largest from one brood and the smallest from the other. The team dropped me at the field station in the late afternoon and I fed all the young kites. It was a brilliant day in terms of both weather and birds, and on my evening walk to the lake I found two amazing and intricate nests of penduline tits hanging in the willows; there were also garganey, spotted redshank and an osprey at the lake. In the late evening I gave the kites another feed of fish and then had supper before crashing out.

Magnus Sylvén arrived in the morning, and after feeding the 10 young kites, I put them in new cardboard boxes with a soft, grass-and-moss lining. We stowed them in a red transit van for the drive to the airport. A few days earlier, I learned from Steve Rooke that the Nimrod could not land in Sweden but would arrive at Værløse, the nearest Royal Danish Air Force base. We drove north to Helsingborg and caught the car ferry across Øresund, just a 20-minute trip to Denmark. Værløse is about 20 kilometres northwest of Copenhagen, so before long we presented ourselves at the guardroom. It was certainly something unusual for the duty officer, but they dealt with us efficiently by escorting us to the control tower and handing us over to an RAF Air Commodore. He was in charge of a joint NATO exercise with RAF, Danish and German aircraft, and was very interested in the kite project. Not long afterwards, the Kinloss Nimrod came down the runway, much to the consternation of the joint controller, a senior German Air Force officer. His 'What is that plane doing here?' was answered quite simply by the Air Commodore: 'Oh, it's dropping by to collect Roy and the red kites.' His opposite number was dumbfounded.

In no time at all the crew loaded the kites on board and I said goodbye to Magnus, thanking him most sincerely for getting us started on what would prove to be an amazing and incredibly successful reintroduction project. Yet again, I had met the right person at the right time for the right project: he was to become a friend for life. The Nimrod was soon racing down the runway and heading across the North Sea to RAF Kinloss, carrying my precious cargo of 10 young kites along with all the official licences. We were met there by Don Gordon of HM Customs and Excise in Elgin and Mr Dundas, the government veterinarian, as well as my colleagues Colin Crooke and Dee Doody. We took six of the young red kites to the hacking cages on the Black Isle, while the remaining four were collected by Richard Porter of the RSPB and Mike Pienkowski of the Nature Conservancy for the long drive to the southern release site in the Chilterns.

Our birds settled in very well, with Dee giving them regular feeds of rabbit and fish chunks. They were inspected again by the local veterinarian on 14 June. The kites were thriving in the hacking cages. We had decided that we should fit VHF radio transmitters to the young birds so that we could track them on their first travels. As well as being ringed, they were also fitted with orange, numbered PVC wing-tags. On 18 July we were ready to carry out the final check before release; Ken Smith travelled up from RSPB headquarters with Robert Kenward, who had made the tiny transmitters to fit snugly on a central tail feather. We started at 1 pm, checking each bird, measuring and weighing them and fitting the tags. The latter was a fiddly job as it involved sewing them on through the base of the tail feather and holding the thin aerial along the feather shaft with dental floss and superglue. It struck me that in future the transmitters should be fitted with a tiny plastic tube so that we could slip the transmitter onto each tail feather without the sewing procedure, which was tricky and could weaken the feather. It became standard practice. We were finished just after 4 pm and, looking through the peepholes in the hacking cages, we could see that they had all settled down as though nothing had happened. The next morning, Mr Dundas, the government veterinarian, arrived to check the kites again and was completely satisfied that they could be released, giving us official clearance to do so. Joanna Buchan of the BBC joined us as well, to do a news report.

On 20 July 1989 the first three kites were released by Colin Crooke and Dee Doody. We had decided that we would release them in a way that gave them absolute freedom from humans, and my idea was that the front of one of the compartments would be slowly lowered before dawn so that, once it was daylight, the kites could fly free in their own time. All three had flown out by 6.30 am and had gone round to perch in high trees nearby. The first stage was underway. I was unable to see the rest of the releases in Scotland because I was leading a wildlife group to Kenya, but before catching the evening flight to Nairobi, I visited the English release site. Ian Evans, the project officer, kindly collected me from Heathrow and drove me to the Chilterns. The Worsley estate proved to be a superb location, a patchwork of fields and mature beech woods, pheasant shooting country. Ian showed me the young that, like ours, were ready to fly. The design of the cages was different and I thought the young were a bit too used to people, but it was great to see the kites I had collected in Sweden.

With kites exploring the Black Isle and surrounding countryside, public interest really blossomed. Farmer friends phoned in with news of a kite on their land and to ask what they might do to help. Our reply was to shoot a rabbit and place the carcass out on one of their fields for the kites to eat. Soon the birds were learning to live beside the local buzzards. Often the latter ate most of the food, but it was amazing how even a rabbit's foot would sustain the kites. On 16 August, after tracking one of the kites at Glenferness, across the firth, I spoke about them from the Inverness BBC studio on Jimmy Macgregor's popular *Gathering* radio show. The kite project was proving very popular. On 22 September it was a pleasure to take Peter Davis, the great expert on Welsh red kites, to see them, along with his wife Angela.

This was a time for laying down lasting memories, such as on 16 October, when I went to the Munlochy Bay lay-by near our office. There I had superb views of two kites swirling in mid-air, chasing a common buzzard, and then being chased in turn by two jackdaws and a crow. They perched in oak trees before heading for the east side woods, where in later years a pair was to nest. It was the first time that I really forgot that they had been absent for nearly a century. On 10 November, I gave a lecture on the red kite reintroduction – the first of many such talks – at the Scottish Ornithologists' Club's annual conference. The year's work was discussed at our red kite meeting at the RSPB headquarters in Bedfordshire on 27 November and plans for the next year were agreed.

For all the excitement, though, 1989 finished in a distressing manner. In December, one of our kites was found dead on the Rosehaugh estate, just 10 kilometres from the release site, killed by a poisoned bait. At that time the land was owned by the Eagle Star insurance company, which had helped so marvellously with the sea eagle project. Some birdwatchers who heard about the incident threatened to move their business away from Eagle Star, and it was clearly an embarrassment to them, for we were given full access to the land and the keeper was replaced.

Meanwhile, other keepers were getting in touch with news of wandering kites. On 17 March, for example, I called to see Tom Robertson, the head keeper at the Altyre estate, near Forres, who the day before had seen one of the wing-tagged kites in the fields near his house. We walked through the fields with no luck, which was not surprising as by this stage they were very mobile.

On 5 March 1990 I had written to the station commander at RAF Kinloss, Group Captain Gould, requesting logistical support, which he promised; I also mentioned that Dee Doody had been commissioned to do a red kite painting for 201 Squadron as a memento of our joint work on the project. Three weeks later, Steve Rooke brought a group from the squadron over from RAF Kinloss to discuss the summer programme and to view the hacking cages and release site: we also treated them to a nice lunch in the pub at Munlochy. All was set for our next collection. Colin and I had decided that, because we were to release 20 young in 1990 and in the following years, we should have two release sites, some distance apart. We kept the excellent site at Teandore Farm, and one of our friends offered a very good site on his small property above the Fairy Glen at Rosemarkie, 14 kilometres to the northeast. Andy Knight took on the red kite post and, once he had arrived, he worked with Colin and Peter Fennel, the landowner, to build hacking cages on the hill top.

Soon, all the paperwork had been sorted and we were ready to import 20 new red kites from Sweden. Steve had been working hard to see if the Nimrod could land at the civilian airport at Malmö-Sturup, which would make transit much easier. On 10 June we heard, from Sweden, that it might be possible, but that the Nimrod must be parked out of view: Sweden, being a neutral country, did not allow nuclear-capable planes, boats or vehicles to visit its territory. There would be no welcoming dignitaries, no publicity and no landing fee, but the crew would be allowed to make a quick visit to the café and duty-free! In Sweden, Johnny Karlsson of Lund University was our main contact, as Magnus had moved on to WWF. I flew again by SAS flight from Aberdeen to Copenhagen via Stavanger. After catching the ferry and bus to Malmö, I was collected by Ulf and Per-Olov. It was great to catch up with them while having dinner with Per-Olov, his wife and children, and to discuss our plans; I spent the night at Ulf's.

The next day was a beautiful, sunny one and we were ready at 7 am. The three of us drove in the van to Stensoffa field station, where I left my gear. The first nest was in the pinewood at Silvåkra, where the mosquitoes were awful. There were three young and I collected the two largest ones; Ulf reported that the nest contained the remains of fish, black-headed gull, rook and brown hare. The next was in a Scots pine on the army ranges; the chicks were too small, but below the tree among the litter which kites collect and take to their nests, I found a brand-new red Swedish Army knife in its sheath, perfect for me to cut up food for the kites. Next was a new nest at Sommarbo, where I took two from the brood of three large young, with the adults circling overhead being mobbed by a hobby while a golden oriole whistled melodiously nearby. We then visited the kestrel wood, where Ulf used the pole-ladder to collect the two youngest chicks. We had done well, with eight young from four nests, so we headed back to the field station and settled them in their boxes in the lab room.

The farmers were cutting hay when we set off again in the afternoon, and adult kites were following the tractors, searching the fields for food such as small rodents and amphibians. There were four big young in the first nest in a big beech tree, so I collected the two smallest ones. Ulf had a really difficult climb at the next big beech tree and the top of his pole was damaged, so after Per-Olov ringed two young and I had collected the smallest, we called it a day. A local garage mended the pole and the team dropped me at Stensoffa. As the year before, my first task was to feed the kites before I fed myself and had an evening walk to the lake. From the bird tower I enjoyed watching 10 pairs of nesting black terns as well as other water birds and waders. Back at the field station I had a worried phone call from Steve Rooke: the Ministry of Defence in London were being difficult about the Nimrod visiting Sweden and wanted confirmation from the Swedish Government. I discussed the problem by phone with Johnny Karlsson and he promised to do what he could.

On 12 June the thrush nightingales woke me with their singing at 4 am. There was mist hanging over the lake and the eastern sky was red. I was soon cutting up food for the kites, then made my own breakfast and was ready to set out for another day's kite ringing with Per-Olov and Ulf. We headed for Ystad and started to visit kite nests, often finding a fledged raven's nest in the same wood: ravens were increasing in numbers, just like the kites. We had our lunch overlooking a rubbish dump, with half a dozen kites to watch, as well as crows and gulls; in the hour, about 20 different kites came in, some of them non-breeding one-year-olds. Several of the kite nests we visited did not have suitable young and, very disappointingly, a nest we checked on our way back to Stensoffa had been illegally shot out, with one dead young under the nest and a slightly injured one in the nest, which Ulf considered would survive. Our return to the field station was not as successful, then, as that of the day before. I had phone messages from Scotland, and Johnny turned up to discuss the plane situation; he had written on our behalf to the

Board of Civil Aviation in Stockholm. I fed the kites, which by now was a big job, with 16 young, before taking an evening stroll to the bird tower, where I enjoyed listening to the fantastic trumpeting of cranes in the marsh.

We had four chicks to collect on 13 June and set off at 8 am, visiting nests on the way to Fyledalen. The first nest was 20 metres up a very high beech tree in a private wood of 300-year-old oak and beech trees; this was where we had found four young the year before, but this year one egg had not hatched, so I collected the two smallest chicks. We lunched with a view across Fyledalen, where we watched a first-year golden eagle hunting along the hillside. I was very interested to see it, because it was the young of one of the pairs now breeding in lowland Skåne. That's the type of rich habitat where eagles should breed in the UK, not just in the uplands, to which they have been pushed by long-term persecution. After lunch we collected two more young from two nests between Norrevang and Sommarbo and reached our total, thanks to the great effort put in by Per-Olof and Ulf. In the evening it took me 90 minutes to feed all the young kites.

Next morning, after feeding all 20 kites, I drove to Johnny's house to regroup, for it looked like we might have to transit to Denmark again. Magnus and Johnny were working their contacts in Stockholm to clear the permissions to land at Sturup. The problem was that the Swedes were happy for the flight to arrive, but London needed more reassurance that there would be no landing fees and that the flight would be welcome. It made for a very tense morning, with phone calls between Scotland, Lund and London. We heard that it was a definite 'no' from London, but then Magnus and Johnny managed to get the Swedish Defence Ministry to phone London to reassure them on the landing fee front. Several ministers in the government also extended official invitations, and the hurdle was cleared. The local veterinarian arrived to examine all the kites and issued me with their official export papers.

It was a real relief when Johnny arrived and, as an enjoyable break, took me across country to visit the white stork reintroduction project. We met the sheep farmer who was running it, having started in March 1989. Three pairs had nests inside the cages, holding large young, while a single pair had a nest with young on top of the aviaries. I enjoyed a nice, peaceful dinner at Johnny's and then headed back to the field station to feed the young kites. I was exhausted.

On 15 June I put new bedding in each kite's box and fed them before doing an interview down the line with the BBC's *Today* programme in London at 7.40 am. Per-Olov, Ulf and Johnny helped me pack everything, and we arrived at Malmö-Sturup at 10.40 am to find 20 journalists and photographers, as well as Swedish TV, waiting for us. It was a big story. We were escorted through to the Terminal 2 apron and I took out two birds for interviews and photographs. We then passed through customs and had our paperwork checked just as the Nimrod arrived, bang on time. Very heavy rain was blotting out the runway, so the plane did a couple of circuits before landing, and by the time they landed I was very relieved and happy to see Steve Rooke and his crew. The kites were put on board and the crew had time for the promised trip to the café for coffee and cakes and the duty-free shop.

Two hours later we were at RAF Kinloss, to be met by a crowd of media, including the team from *Blue Peter*. The RAF crew, of course, were delighted to be awarded Blue Peter badges. Colin and Andy had arranged for half the kites to go to each hacking site. It was another very long day, but had turned out as well as it did because of the huge support from friends and

Flight Lieutenant Steve Rooke (RAF Kinloss) with the author and a young red kite at Malmö airport, June 1990.

colleagues in both countries. That's why I love these projects so much.

Andy Knight had a busy time feeding the full complement of young kites every day. Once we were sure they were feeding well, Andy put fresh food – a mix of dead rabbit, fish and road-killed birds – through the feeding sleeves onto their nests each evening. One young had problems with its leg joints and, despite treatment by John Easton, our local veterinarian, it died. It was a chick from a nest near the rubbish dump and had clearly not been receiving nutritious food from its parents. I decided we would collect no further young from nests near the dump. The remaining 19 birds were cleared through quarantine on 23 July, when they were ringed, measured and fitted with tail radios and blue wing-tags. Blood samples were also collected for screening and future DNA fingerprinting. We were extremely pleased with their condition, so they were released between 25 July and

2 August. The young kites stayed near the hacking cages for a longer period than in 1989, probably due to the larger numbers, but despite continued feeding at the hacking cages, they quickly found their own food successfully on the Black Isle and started to range.

My assistant, Colin Crooke, accompanied me on the 1991 collection trip to Sweden. The RAF station commander had obtained agreement to land the Nimrod at Malmö, and Steve

Three young red kites in hacking cages prior to release on Black Isle.

Rooke sorted out all the arrangements. We arrived by our usual route to Lund to stay the night with Ulf and have supper with Per-Olov's family. On 17 June we set off early to visit kite nests; our first was in the pinewood on the ranges, where we collected two young from a brood of three. A new nest, near Norrevang, also provided two from a brood of three. Leif Hansson joined us and we visited nests all morning, several of them new ones found earlier by Per-Olof. The population was clearly increasing. One was in a small pine that I could climb, but I found just one chick for ringing; food remains in the nest were eel and pheasant leg. We lunched at the white stork project, where this year four nests were outside the aviary, but heavy rain had killed three broods. We carried on checking nests between Blentarp and Vressel, with a tally of 10 young collected by the end of the day. It was an easier job feeding the young with two of us.

The next day, 18 June, was a complete washout, with heavy rain all day preventing tree climbing and nest visiting. Fortunately the weather had improved by the following morning and we could do fieldwork, even in rain showers, although climbing big beech trees was difficult for Ulf. Part of the time we had a Swedish TV crew with us but managed to collect the remaining young during the day's kite ringing. Everything followed the usual routine and the Nimrod got us home at 3.45 pm. Duncan Orr-Ewing had taken on the kite project, so after placing some of the young at Teandore, Colin and Duncan took the others to Rosemarkie. We tagged and checked all the young kites on 26 July, and the dawn releases started at Teandore on the 27th and Rosemarkie on the 28th. It had been another successful year, and it was even beginning to feel like routine.

It was very clear to me, though, that the success of our visits to collect young kites in Skåne was due to Per-Olof's outstanding knowledge of the location of nests. In summer, when the

Duncan Orr-Ewing and the author radio tagging a young red kite before release. (*Photo by Duncan Orr-Ewing*)

beech woods were in full leaf, we would follow him until he pointed to a specific tree and said, 'There's a new nest up there.' Peering up through the dense green foliage, we could just about see it, but Per-Olof had, of course, found it in winter when the 'grey skeleton of the beech' had revealed its secret. To see this essential fieldwork in action, he invited me to visit Sweden to join him on his recce weekend in the first week of April 1992. I arrived at Lund in the evening and next morning we were off to the field by 7 am to visit the nesting areas I knew from my summer visits. Now, the great beech trees stood leafless and quiet, the landscape still in its winter mantle, with grey skies and early drizzle. We'd see a kite over a wood and I would follow Per-Olov to check the nest: some were the same as in previous years, while others had moved to a new tree. It was good to see the adult pairs circling overhead and to find that the ravens were at home with young in their tree nests. In this season, the black woodpeckers were vocal and easily seen, unlike in summer, while small migrant flocks of white wagtails were arriving on migration. By evening, Per-Olof had taken me to 26 nest sites and we had seen over 50 adults.

We had another early start on Sunday in the woods near the field station, quickly locating the adult pairs sitting in the bare woods near their nests. That was all that Per-Olof needed – to know which nest was going to be used – and then on he'd go to the next site. We crossed farmland to get to two nests; the fields were just cold, heavy, black soil, in contrast with the crops and flowers of summer.

At the dump we counted about 25 kites as well as a young sea eagle and two rough-legged buzzards. Walking through the bare woods there, we found that six nests were already taken by kite pairs, an increase of one on the year before. After Krageholm, we visited a wood of big beeches and watched a pair of black woodpeckers – jet-black plumage with brilliant red crowns and an impressive white bill against the grey bark – as they excavated holes in three trees. And then came two flocks of 19 and 26 cranes in formation, wheeling overhead, calling loudly as they returned to Sweden for the spring. Our tally that day was 45 nest sites and over 150 red kites. I now understood how Per-Olof could take us unerringly through the summer green forests to occupied nests. It was down to dedicated fieldwork in all weathers, the true basis of a field ornithologist's year.

In May we were delighted to host Per-Olof and Ulf at my home near Loch Garten, in the Scottish Highlands, and show them 'their' kites, including the first nesting attempt on the Black Isle. A month later I was walking in the woods of Skåne with Per-Olof and Ulf to collect the new cohort of young kites, and what a difference a few months made. The countryside was very dry after a heat wave and some broods of young kites were too big to collect for our project. By evening, we had 10 young safely in the field station and they needed feeding. The next morning, Johnny joined us at the first nest near Simonstorp, where the local hunter took us into a 200-hectare hunting reserve for wild boar, fallow and red deer. It was a superb nest in a beech tree on a hill slope with three excellent young, and I collected two. By the time we reached the white stork farm we had collected eight young. We had coffee with Karin and she told us that 16 young storks had been reared but that wild food was scarce due to the hot spring. Our last kite that day was from a nest in a cherry tree by a wheat field, and then on we went to the field station to feed all 21 of them. The next day we completed the collection at 24 young and, after feeding them all, drove to Johnny's house for an excellent chicken supper followed by strawberries; and the Swedish football team beat England 2–1. The next day, the RAF Nimrod arrived exactly on time at midday, Steve having organised everything again, and by late afternoon all the kites were installed in the hacking cages on the Black Isle.

The last year of collecting was in 1993, and although the main task was the same, the variations created interest. This year we had a film crew with us when Per-Olov and Ulf took me to the kite nests. At the first nest, the young were too small, but the next had two young and I collected one while doing an interview to camera. We had collected nine young by the time we reached the white stork farm at Hemnestorp. That nest had two dud eggs and just one young, then on we went to the nest in a low birch, where I collected the two smallest ones. After feeding the young back at base, I walked to the lake and watched bearded tits, black terns and marsh harriers. The next day we got off to a good start and there were suitable young at six nests, but heavy rain showers later curtailed our fieldwork. Liz Ingelog, of WWF

Sweden, joined us, as well as a Swedish film crew. On the way back we stopped at Fyledalen; I had become very fond of this beautiful valley. We watched the adult pair of breeding golden eagles and, in a small reserve, three pairs of gorgeously plumaged red-necked grebes. The 16th was another wet day, although we did reach 23 young, meaning we had just one to go. In the evening we had a nice party at the field station with our Swedish colleagues and friends, which gave us a chance to thank everyone for their superb support.

The next day was also too wet in the morning for the BBC crew to do any filming, so we took time off to visit Falsterbo and its lighthouse bird observatory. There, over coffee, I caught up with Lennart and Karin, the bird wardens, and saw some interesting birds including nesting Kentish plovers. At noon, the rain ceased and we filmed the collection of the last young kite. On 18 June Colin and I fed the kites in the early morning and then boxed them all up with Per-Olov before heading to Malmö airport. The Nimrod arrived after midday and it was really excellent to see that the RAF Kinloss Station Commander, Group Captain Andy Neal, had come on the last flight with Steve and the crew. It was a chance for me to convey our huge appreciation to RAF Kinloss and to the squadron. Without them, I'm not sure the project would ever have got started. Our two-hour flight back home was enlivened by the Nimrod being hit by lightning while our kites 'slept' in their boxes. Duncan and Brian Etheridge (himself originally a Nimrod crew member before working for the RSPB) arrived to collect the kites.

Over the five years we imported 98 red kites from Skåne and were proud that we had reared and released 93 of them, while four had flown free in England. That's such a simple statement, but in fact it involved a huge amount of hard work and dedication. In Sweden, Magnus Sylvén, Per-Olov Andersson, Johnny

Karlsson, Ulf Sandnes and Nils Kjellén were magnificent, and without Per-Olov's high-quality fieldwork, and Ulf's tree climbing prowess, we could not have carried out the project. Our official contacts in the Swedish and UK government bodies and veterinary officers were excellent. Flight Lieutenant Steve Rooke, the Nimrod crews of 201 Squadron, and the Commanding Officers of RAF Kinloss provided the rapid transport necessary for such a venture and we appreciated their support and friendships. The other major – and crucial – task was to rear and release the red kites in Scotland and we were very fortunate that the kite project officers of the RSPB – Dee Doody, Andy Knight and Duncan Orr-Ewing, guided by Colin Crooke – carried this out to a very high standard. It's a tough and demanding job to collect and cut up dead rabbits, fish and birds for the daily feeding of the kites. It's a messy task but, when it leads to results like ours, it's worth it.

Meanwhile, from 1990 onwards, the English red kite team established a very good relationship with ornithologists in northern Spain, and over five years imported 90 young red kites to release in the Chilterns. The kites from Spain were likely to be sedentary, unlike the partially migratory ones of Sweden, and at the beginning I was not certain that this was a wise choice of donor stock. In retrospect, I was wrong, because the Chilterns kites basically stayed where they were reared and the population grew much faster than ours in the Highlands. The northern ones suffered higher mortality when they were juveniles, some got lost – one even ended up in Iceland (but was sent back) – and, disappointingly, too many of them visited shooting estates on their wanderings, which were illegally using poisoned baits. Some practices had not changed since the Reverend Forsyth's time.

It was tremendously thrilling that the project teams could start new releases as originally planned, in Northamptonshire in the south and Perthshire in the north. Duncan Orr-Ewing, the kite officer in the early days on the Black Isle, managed the Scottish second release. He established a link with eastern Germany and, over five years, imported 103 young red kites. I remember this population well, because Dr Michael Stubbe of Halle University invited me to visit the famous red kite breeding grounds of the Hakel on 26 September 1995. It is a 1,300-hectare deciduous wood on the east German plain, made up of oak, beech and linden trees; when I visited, it still had an interesting raptor population including red kites, black kites, four pairs of lesser-spotted eagles, three pairs of honey buzzards and a newly

The first red kite chick in Scotland of the new era – he went on to be a regular breeder on the Black Isle. The nest is well stocked with dead rabbits, including a black one.

arrived pair of booted eagles. In earlier, Communist times it had hosted over 100 pairs of red kites, though, due to an extremely plentiful food supply of golden hamster. The traditional agriculture prevailing then was beneficial to the rodents, even though in some years the authorities might kill 400,000 hamsters to reduce agricultural damage. The kites thrived nonetheless, but after reunification and the arrival of intensive farming methods from western Germany, the hamster populations collapsed and the kites fell to a dozen or so pairs. That day, Michael showed me nests of the two eagle species and a goshawk nest; we saw about 40 red kites and noted signs of large numbers of common voles around the wood edges.

Both the new populations survived, so in Scotland Duncan organised a new release area in Dumfries-shire, with young kites from the Black Isle and the Chilterns, while in England the new site was in Yorkshire. Subsequent releases of red kites were carried out in Tyneside and Aberdeenshire. The project plan we devised in 1987 had been carried out successfully, and the red kite was once again a common sight in many parts of England. Successful translocations were carried out in Ireland and Northern Ireland, with smaller translocations elsewhere, and nowadays it's thought there are well over four thousand or so pairs in the British Isles. At the same time, the Welsh red kites were thriving, especially when they spread to biologically richer countryside, and some joined up with the English birds. I remember the first time this really came home to me. I was driving down the M40 and saw a flock of 20 red kites swirling over a road as it cut through the northern Chilterns. More recently I've seen them over the M25 as I've been driving to Heathrow Terminal 5 to catch a plane home. This is what I'd always hoped for: that the red kite should be part of our whole countryside, not a denizen of remote valleys. The way that they have started to live among us in towns and villages reminds us that those old observations about London in the time of King Henry VIII were true, and need not be consigned to history. It has been the most amazingly successful project.

Red squirrel in Moray – at a donor site for the Wester Ross translocation.

# 4

# BUREAUCRACY

## *Obstacles along the road to reintroduction*

Sometimes I found it frustrating that pro-active wildlife management would often be delayed by rules, regulations and prejudices that seemed, to me, not to be in the best interests of the wildlife we were tryng to conserve. Too often, people would refer to the dangers to nature of grey squirrels or Canada geese, while ignoring the fact that they were clearly unfortunate examples of non-native introductions. Our projects involved reintroducing or translocating native species to places where they used to live, but had been eliminated. This chapter involves two species which became caught up in bureaucratic hurdles.

## Ospreys to Rutland Water

I find it hard to believe, with hindsight, that there was ever a time when I did not see osprey translocation as part of the picture in the UK. Yet there I was, attending the Raptor Research Foundation annual conference in 1981 in Montreal, not realising where the exchange of ideas might lead me. The focus was on bald eagles and ospreys; both species were slowly increasing in the United States after the catastrophe of water-borne pesticides and contaminated fish. I had been invited to attend to give a talk on osprey conservation in Scotland. It was an inspiring three-day meeting, with many like-minded raptor workers from North America and a few from Europe, but we felt like poor relations: the venue was an expensive, skyscraper hotel, and a group of us from Europe had managed to cram into one large bedroom to make our money last. Meals in the hotel were beyond us, so we ate from the fast-food outlet across the road. The important thing, though, was that we talked and learned about the conservation of our favourite birds. I found some of the reintroduction projects carried out by the Americans particularly inspiring, but still didn't see that what they were doing could be done by us at home.

In Scotland in 1981 there were 26 pairs of ospreys, and the main conservation work focused on protecting those pairs from egg thieves and building artificial nests to spread the population and increase breeding success. A few years later, I visited Rutland Water nature reserve at the invitation of the manager, Tim Appleton, who was to become a lifelong friend. He wondered if it would be possible to attract passing ospreys to stop and breed, so I helped build a nest in a big ash tree by the water's edge. No bird ever used it, but in 1994, on 15 May, a female turned up, followed by a male six days later. They did not nest but spent the summer at and around the reservoir. Tim got back in touch to see what could be done to encourage them further. I advised him that they were probably immature birds that would head north next summer, but I had a plan that was worth a try.

By that time there were 95 pairs nesting in Scotland, and we had become very proficient at encouraging ospreys to use nests that we had built in new areas. I recommended that we tried building more nests, on poles, around the reserve. I sent details of the nest platforms for the reserve staff to construct, with Tim agreeing to scrounge five disused electricity poles.

On 16 March 1995 we started with a recce to find five good sites on the reserve, before the team put out the poles and materials. At midday the East Midlands Electricity Board's guys arrived with a digger, free of charge, and we erected the first pole in a lagoon. I climbed the pole and Tim's crew hauled bundles of sticks and bags of moss, dried grass and leaf mould up to me. It took an hour to get the first nest built. Next we built another on a very tall pole on Brown's Island, with the BBC and the press recording and photographing the event. It had been a lovely, sunny day, so we decided to work with the digger driver and get the remaining three poles erected, then build the nests in the morning. That evening I was very tired, but happy to give an illustrated lecture on ospreys to over 100 people in Oakham.

Next morning, we started on the third nest, but by now the weather had changed, with increasingly strong westerly winds, rain and even hail showers. Nevertheless, we pressed on and completed the first nest just as Tim returned with a photographer from *The Times* and Stephen Bolt, the Anglian Water senior ecologist; the company owned and managed the reservoir. The next pole was the most southerly and tallest but, with the crew's help, I built a really good, big nest while standing on the top rung of the ladder. I remember standing at nest height, looking around the reserve and asking Tim about the quantities and species of fish in the reservoir. We had lunch at Tim's cottage and I broached the possibility of reintroducing young ospreys from Scotland, rather than waiting for them to arrive. Tim and Stephen were enthusiastic and we had a great discussion. By the time we got to the last nest, the wind had risen to a gale, we found it difficult to get the ladder extended, and once tied into the top of the pole my ropes blew out horizontally. Somehow I managed to get two bundles of sticks tied into the nest and added a bag of grass before baling out as rain, hail, thunder and lightning bore down on us. We were wet and exhausted but the job was done, and after hot tea and a bite to eat at Tim's, I drove the 705 kilometres north to arrive home at 11.30 pm. How I loved those days of tough work with great colleagues.

That spring, there was no sign of the male, and while the female did visit, she did not linger; they were probably back in Scotland. We had, though, started to prepare a case for a reintroduction, and I wrote a draft report for Anglian Water and Tim on how we would carry out a reintroduction of Scottish ospreys to Rutland Water, assembled the scientific evidence required for a feasibility report to accompany a licence application and, for Anglian Water, drew up an idea of the budget required to run the project for five years. I was able to submit this report on 22 May 1995. By this time, the idea was being talked about and I quickly realised that there were differing views at either end of the country. There was, in fact, outright opposition from some quarters, including from the RSPB at Loch Garten, where the wardens said that if ospreys started to nest in England, fewer people would visit the Loch Garten osprey visitor centre. This would be detrimental to income and to membership promotion. I did not agree, as I was sure that osprey enthusiasts would visit all future osprey centres. For me, it was important that rare birds should be as well distributed throughout the country as habitats and food allowed, rather than kept rare as special kudos for a reserve or even a particular county or country. As wildlife managers, we should endeavour to encourage the recovery of a species in as much of its lost range as possible.

In late October we had a meeting of the red kite committee at the headquarters of English Nature, originally the Nature Conservancy Council, including the RSPB and country conservation agencies. It was an encouraging meeting, for the red kite project was turning out to be very successful. That evening I went to stay with Ian Newton and his wife; Ian has always been a powerful advocate for reintroducing and translocating raptors to encourage range expansion of species like white-tailed eagles, goshawks and red kites. Throughout my work with these projects he has always been one of their staunchest supporters. That's very important, because Ian was and is one of the most respected raptor ornithologists, not just in the British Isles, but in the world.

Next morning, 24 October, we held the inaugural meeting at Rutland Water of a putative steering group for the osprey project. It was hosted by Tim Appleton, Stephen Bolt and myself, with Greg Mudge from Scottish Natural Heritage, Phil Grice and Ian Carter from English Nature, Michael Jeeves and Hugh Dixon from the Leicestershire and Rutland Wildlife Trust and Nicola Crockford from the RSPB. It was a good meeting but we had a job to persuade some interests round the table. There were, to my mind, too many unnecessary worries about what could go wrong. Nevertheless, there was agreement that we could go ahead and recognition by us that we would need to fulfil all the necessary International Union for Conservation of Nature criteria, although some might say that the IUCN guidelines, which were meant as a guide to good practice, to ensure successful reintroductions, were being rigidly interpreted by some to make projects more difficult to carry out. We had an encouraging visit to the osprey nest platforms, erected early in the year, which helped with detailed discussions on the project.

That winter we worked on our feasibility document, explaining all the processes we would undertake, from collecting one young from broods of two or three, to their care in Scotland and transportation to Rutland Water, their subsequent time in specially built hacking cages, followed by their release back into the wild. We needed to explain that we would not be damaging the donor population, and for this we were greatly helped by Dr Mick Marquiss, who examined the breeding data and provided the scientific evidence to show that the taking of up to 12 young a year would have no detrimental impact on the future of the Scottish population. In Scotland, landowners and the Forestry Commission were immediately supportive of the project and, if we were to get a licence, would allow us to collect young on their lands. We were not allowed to take young from designated sites such as special protection areas for birds, and we decided to avoid nature reserves. We proved that the original osprey distribution in Britain was widespread and so we were restoring the species to its original range. We also demonstrated that Rutland Water and surrounding wetlands could host a population in the future of about a dozen breeding pairs. I had also received advice and information from experienced birders, whom I had met at the Montreal conference, and was given the latest information on osprey translocation projects in North America. Between 1979 and 1990, osprey reintroduction projects in 12 states had moved over 900 nestlings. Several people invited us to come and see their work.

In Scotland I was working towards a potential start date in the spring; and in late February, at the annual conference of the Scottish raptor study groups, I gave a talk on our proposed osprey reintroduction to central England. There was a lot of interest and support, especially following on from the success of the red kite reintroduction to Scotland. We submitted the licence application to Scottish Natural Heritage and were asked to supply supporting evidence on various issues. At this time I was on the main board of Scottish Natural Heritage so could answer queries from senior staff when we met at our various headquarters. Nevertheless, some members of staff had misgivings about the possibility of things going wrong, so a subcommittee of three professors from the board of SNH were tasked with examining the proposal. We needed to be patient and wait, hoping for a licence in the end.

Twenty-four years later, it's hard to believe the antipathy and opposition in some quarters towards translocations like the osprey proposal. Scientists argued that our approach was not scientific enough, often believing that science was the sole arbiter. Others suggested that everywhere in England should be examined for the reintroduction rather than, as they saw it, just plumping for Rutland Water. To me, that's a sure way of delaying progress. Some said the only reason Anglian Water and Tim Appleton wished to have ospreys was to make money. The temptation to complicate the process was a worry, for those efforts produce reams of paper but little else. The RSPB were initially against the reintroduction but then decided to be neutral; not in favour, but not actively against. Now the Society is much more proactive – witness their excellent recovery project on cirl buntings in Devon and Cornwall and the head-starting (the hatching of eggs and subsequent release of young) of black-tailed godwits in the Ouse Washes. This new century has witnessed a recognition that in a world of catastrophic declines in wildlife, we need to be bold and innovative. Back then, though, we had a tense wait to see if we would be given the green light or turned down flat.

Meanwhile we took up an offer from an American ornithologist, Mark Martell, to come and see his osprey reintroduction work in Minnesota, and particularly around the Twin Cities. Anglian Water kindly, and importantly, funded a study trip by Tim Appleton, Hugh Dixon and myself. On 26 April we flew by Northwest Airlines to Minneapolis, where Mark collected us. In no time he had taken us to the first of his release areas at Henneken Park. We walked to the lakeshore, with killdeer plovers and double-crested cormorants reminding us we were birding in the US, and examined the nesting platform erected close to the release site in 2005. Next we visited Fraser Lake, where the hacking site had a backdrop of the skyscraper skyline of Minneapolis – these were truly urban ospreys. The next morning, Mark collected us and we drove northeast to Crex Meadows reserve in Wisconsin, an incredible 13,400-hectare wetland then wakening from winter, with none of the trees in leaf and just the first warblers arriving on migration. We saw that three out of six osprey nests were occupied by single birds just arrived back from South America, while another hunted over the lake. It was a marvellous opportunity to obtain first-hand knowledge from a person who had been involved in running an osprey reintroduction project. We saw several pairs of bald eagles, had superb views of sandhill cranes, heard the lovely calls of three pairs of trumpeter swans and saw a black bear with three new cubs venturing onto the marsh. All that, and coffee with blueberry pie.

On 28 April Mark drove us out west of the city, where we met one of the osprey team leaders, Judy Voigt Englund, who showed us the hack sites and the nesting platforms. Two females were settled on nests and several other pairs floated around, having just arrived back on migration. From a hide at the water's edge we watched a pair on a pole nest platform with the male collecting waterweeds to line the nest. In the distance, on a nest, was a female that had been fitted with a satellite transmitter. The next morning, one of Mark's assistants drove us around a number of osprey nest sites and hack sites north of the city; at one place we admired an osprey breeding on a pole erected in a large garden. We had a great chat over an excellent lunch with the homeowner, Art Hawkins, and his daughter Amy, and as we left he showed me an American robin's nest by his front door, with three beautiful blue eggs. Back at the raptor centre I gave a talk on ospreys and their conservation in the UK, and our proposed reintroduction, which brought our study tour to a close. Before heading home to Scotland, I hired a car so that I could visit Aldo Leopold's shack, where he wrote the ecology classic *Sand County Almanac*. He was, and is, one of my heroes so, for me, this was hallowed ground. I could hear the sandhill cranes calling in the marshes, just as he could when he lived there. As a professor of wildlife management he created a new ethos on the care of ecosystems and the planet. His death fighting a forest fire was a dreadful tragedy not only for him and his family, but for ecology and wildlife management as well. The International Crane Foundation at nearby Baraboo then hosted me for two days, to which I will return in a later chapter.

On 6 June I had a long day travelling from Aberdeen airport to Stansted. Tim collected me and we had a good get-together with Stephen at the Anglian Water offices in Huntingdon, before a day in the field at Rutland Water, planning and checking final arrangements for the arrival of ospreys. Helen Dixon had been appointed osprey project officer, so we were ready. Stephen took me back to Stansted and I headed home, tired at the end of an 18-hour day but feeling very happy; I was even happier on 10 June, when I received the licence from SNH. A week later, Tim and Stephen were with me in the Highlands, and in two half-days I showed them 10 osprey nests in my study areas, which I had been monitoring since late March and some of which would provide donor chicks. The final checks in late June and early July were important for me to judge the age of the young and the numbers in potential donor nests, as my licence restricted me to taking a single young – preferably the youngest, which might be a runt – from broods of three.

The collection of young ospreys started at 7 am on 6 July, when Jo Hayes and I collected a camera crew at Boat of Garten en route for nest A11 near Carrbridge. This was a long-used nest with a special breeding female, which I had colour ringed Green J on the Black Isle in 1991. She had three young. We put a ladder against the nest tree, a dead Scots pine, and Jo went up and brought three young to the ground in a bag, where I ringed and measured them, and collected the smallest. We were started. The next nest was on a farm near Fochabers and, with the farmer, we were able to drive the 4x4 to the bottom of the Scots pine which held the nest. Our big ladder reached high enough, and we found three big chicks – again, I took the smallest. The final nest that day was in a really big tree near Forres, which my friend Bob Moncrieff climbed and we collected the largest of three small chicks. Back at home I had constructed pens holding an artificial nest, into which I placed the young and gave them cut-up fresh rainbow trout. Helen arrived from Rutland in the evening; we ringed and collected young ospreys on the Black Isle and in Sutherland over the next two days and collected the last two young on 9 July. Helen's van from Anglian Water would not start, so the seven young, well fed with fish, settled in their transit boxes lined with hay, were stowed in the back of my car. Helen and I took turns to drive through the night, for I had decided that night-time was the best for transporting the young – it was dark and cooler, with less traffic on the road. We have done the same with our projects ever since.

Tim Appleton, manager of the Rutland Water Nature Reserve, at an elevated hacking cage for young ospreys in 1989 – his faithful collie at the foot of the ladder.

We finally arrived at Fishponds Cottage, Tim's home beside the reservoir, at 6 am to a welcome from staff and volunteers and a big press presence. The young ospreys were now in Helen and the team's care and, to start with, this was very hands-on, as the smaller chicks needed hand feeding with tiny pieces of cut-up fish, while the larger ones could feed themselves. Martyn Aspinall, Tim and the team had carried out the construction of elevated hacking cages, T-perches and feeding nests for after release, and a caravan was fitted with CCTV to the cages. Here, the volunteers maintained a continuous watch and kept a detailed record. Helen was having difficulty getting the youngest runt chicks to improve and, even with the advice and support from the Institute of Zoology in London, we lost two young. Post mortems by Dr Kirkwood concluded that salmonella infection was the cause. On 28 July I drove down to Rutland Water with the remaining young osprey, making a total of eight, which we placed in a cage with a larger one. In the afternoon we examined and weighed the four young on Brown's Island, and carefully fitted a tiny VHF transmitter to the central tail feather in order to monitor them after release.

The first young were released on 30 July, when the front of the cage was lowered using a long cord so the birds could come out in their own time. Over the next two days the remaining ones were released and soon some were perching on a big nest built in a high tree on Lax Hill. I went north on 3 August and was sad to hear later that two more of the runtish chicks had died. In the end, four birds left on migration. One was observed, identified from its colour ring, eating a fish at Pilsey Sands in Sussex on 14 September, while the ring of another was reported from Senegal in February 1997. At least one had reached the correct wintering grounds, although sadly it had died or been killed.

During my years of fieldwork on Scottish ospreys, I had recognised that some parents were always successful and often reared broods of three young, while other pairs usually lost some of their young and finished off with either one or two chicks, or even none. I put this down to the fishing skills of the males – some are excellent fishermen and others are poor at catching enough fish for their family – while in local hierarchies the dominant males always have first choice at fishing locations. In consequence, some nests might hold three big young while others had young of different sizes, due to food shortage, and it was not unusual to find runts. This was exacerbated in long periods of heavy rain, when fishing was difficult and even whole broods would die. In the wild there were clear differences in the ability of adults to rear strong young. In consequence, following the disappointments in 1996, I made a decision to alleviate those problems for our project. Firstly I decided that we would never again collect runt or small chicks. We would choose young of an older age, around six weeks, with a minimum wing length of 320 mm. Young ospreys of this age were capable of feeding themselves with cut-up pieces of fish without the need for hand feeding. I also requested that we could collect single chicks from nests of two or three young, which would give us a greater opportunity to make certain we were only collecting and translocating really good young. These changes greatly increased our ability to rear and release young ospreys successfully, for 55 of 56 young ospreys collected between 1997 and 2001 migrated from the release area, the only loss being a newly fledged young which flew into power lines on a neighbouring estate. The other important change was to move the release cages from the low-lying part of the reserve, where young had got temporarily caught in long vegetation and bushes, up onto Lax Hill. There the cages were in an airy location, with short pasture in front, mature trees with a nest behind and great views of the reservoir.

In early November I held a meeting about beavers with the research staff of SNH in the morning, and then had a good discussion with Andy Douse and Vin Fleming about the difficulties of collecting osprey runts and my suggestions as to how we could address the problems. These were discussed further at the main project team meeting at Rutland Water on 8 November, which reviewed the 1996 season and its problems. Helen had prepared her report on work at Rutland, and Dr Kirkwood spoke about the disease issues. The group agreed with the various changes suggested for 1997. The previous evening I had given a talk on ospreys in Scotland and the reintroduction to about a hundred local members of the Trust in Oakham.

That spring the new hacking cages were completed, and in early July we carried out our osprey ringing and collected 10 strong young. Helen and I drove them down in a hire car and arrived at Rutland Water at 1 am: it was a beautiful, starlit night when we walked with Tim, Martyn and Paul to Lax Hill to place the ospreys in the cages. I was woken by torrential rain in the early morning and we raced to the release site to place sheets on top of the cages, for we knew the young birds would be sodden. Within minutes they were standing up and shaking their feathers, and later in the morning we installed small wooden roofs on the top corners of the cages to give shelter to the artificial nests. This was standard practice from then on. I handed over the two remaining chicks to Tim Appleton and Paul Stammers at a meeting place near Edinburgh on 29 July, and Helen and her team reared and released all of them successfully. Our annual meeting in November was very positive, with Phil Grice and Ian Carter from English Nature and Andy Douse from SNH complimentary about the changes we had made which led to that year's success.

Through to 2001, our osprey year followed a regular course. Starting in late March, I carried out detailed monitoring of all the nests in the north of Scotland. As the numbers increased, more of the nests in some areas were monitored by other members of the Highland Raptor Study Group. In late June I checked out the most likely suitable nests for collecting chicks from land where we had permissions. The important requirement was to know how many chicks were in each nest and how many were of the right age and condition to collect. In the second week of July I went round with my tree climber friends and ringed, colour ringed, weighed and measured the young ospreys, and collected those that were suitable. We only collected really healthy big young. Once collected, they were placed in a box lined with moss in my 4x4 and taken home, where each summer I constructed pens in the garage where they could feed on fresh trout, away from humans, in the quiet. The chicks were driven down overnight to Rutland Water, where Helen and her team carried out the big task of rearing and releasing them, followed by five weeks of placing fresh fish on the top of the cages or on feeding platforms to get them into top condition for migration. Cutting up fish, a mucky and daily task, was one of the essential parts of the project. Finally, in November, we would have our annual meetings to discuss progress and to plan for the following year. It was a wonderful project to carry out with a very good team in Rutland Water and Scotland.

Over the years I'd kept in touch with Mark Martell in Minnesota and other friends from North America and learned about the new and exciting studies of osprey migrations using miniature satellite transmitters, which gave incredible insights into the ospreys' lives. We thought it would be good to study our birds in a similar manner, so in 1998 I came up with a plan to tag some adults and some young in Scotland, as well as sibling young from the same nests which had been translocated to Rutland Water. We were very fortunate that Anglian Water agreed with the idea and offered to fund what were quite expensive bits of kit. I applied for permission to use them to the BTO and to Scottish Natural Heritage. I applied to the French satellite tracking company CLS in Toulouse, for an agreement to use their Argos tracking facilities, and then placed an order with Microwave Telemetry in the US. Mark and others gave me

Jim Gillies (Forestry Commission Ranger) and the author ringing young ospreys in Badenoch and Strathspey, and selecting young for translocation to Rutland Water.

advice on scheduling and fitting the transmitters as well as understanding the data.

Those early transmitters were battery powered and I had to work out a schedule of transmissions to obtain the optimum results for the life of the battery; for example, every other day on migration and every 10 days in winter, then every day again for the spring migration. We decided to track three adult females, three adult males and young from donor nests in Scotland, and also tag young at Rutland Water which had been taken from the same nests in Scotland. It was very exciting to catch the first osprey for satellite tagging. It was on 26 July 1999 and the bird was Green J, the breeding female at her nest near Carrbridge, from where I had collected a young for Rutland in previous years. I caught her in a high net when she dived at a decoy eagle owl near the nest. I carefully fitted the transmitter on her back, beside my Land Rover, and on release she flew to her nest with two young; soon she was feeding them from a fish on the nest. Back in my office, I found I had ringed her as a chick eight years before, north of Inverness. It was a big surprise to me, when she migrated, that she only travelled to winter in a big reservoir in Extremadura in central Spain, and did so for the rest of her long life of 25 years. She never was Africa bound.

The other females wintered in Senegal and in Guinea-Bissau, and the males also headed to favoured wintering sites in West Africa. The young birds had more risky migrations, as they tended to have more southwesterly tracks, and several young ones that we satellite tagged at Rutland ended up near Land's End, which isn't a good place from which to fly to Africa. We already knew that if the winds were extremely strong easterly in southern Europe and North Africa, ospreys could be blown out to sea and lost, but we proved that the translocated young wintered in the same areas of West Africa as those from

Scotland, despite the 600-kilometre translocation north to south from Scotland; they behaved like their siblings back home. Because the radios were battery powered they did not last long enough for us to study the return of the young, although one female amazed us by returning as a one-year-old and breeding at two for just one year, which was very unusual. These results were enhanced by later, more sophisticated, GPS transmitters, but we became worried that adding satellite transmitters to young translocated birds did not justify the gains and that it was best to reduce any possible impacts.

That year, 1999, also saw the first return of a translocated osprey – a two-year-old male – to Rutland Water on 29 May, and a second male, also a 1997 release, on 14 June. These two returned in 2000, and occasionally one attracted a ringed female, probably from Scotland. Later, two more males from the 1998 release arrived. Things were looking good and looked even better when the farmer reported a nest. When Tim and Helen went to look, they found a beautifully constructed nest, built by one of the three-year-old males, the second to arrive in 1999.

Everyone was very thrilled at the start of spring 2001. The two older males returned and, not long after, one attracted an unringed female to the nest found by Tim. This became known as site B. The female laid eggs and the pair started incubation. On 7 June I was excited as I drove down to Rutland Water; it was an honour to walk with Tim, Helen and Ian Newton to the makeshift hide in the edge of a wood to view the big nest on a tall, stag-headed oak at the other side of the field. The male was eating a fish and the female incubating low down in the nest; the male flew off and the female briefly fed a chick from the edge of the nest before settling down. It was so thrilling: there were young at last, five years after we had started the project. These were the first young osprey in central England for more than

150 years, and this was the first successful reintroduction of ospreys in Europe. We celebrated that evening at Tim Appleton's house with fish and chips and a good wine.

Early next morning we went to the visitor centre and watched another female on a nest platform on Brown's Island being displayed at by a male high in the sky. And these were not the only ones. Back at Tim's cottage, Andy Brown, the new Anglian Water ecologist, had arrived and so we set off with the osprey team in Tim's Land Rover to nest site B. I had brought my osprey nest-checking mirror-poles from Scotland: using 10 aluminium sections jointed together, with a small mirror on the top, we were able to look into the nest and saw one chick and two eggs. We quickly withdrew and, before we had gone far, the female came back, just as ospreys do in Scotland. Sadly the other two eggs did not hatch, and my view was that this was an old female, past breeding age, which had been turfed out of its nest in Scotland by a young pair. On 14 July the female chick was taken briefly from the nest for weighing, measuring and ringing by Tim Appleton, and he added a red colour ring 13, which might have been an unfortunate choice: after migrating, she was never seen again, although she may have bred somewhere in a secret location. That summer, a pair of adults from Scotland reared a single young in the Lake District, which marked the start of colonisation in northern England. Sadly, some senior conservationists said this was a 'real' osprey, unlike the young reared at Rutland. This comment had been made before with both the sea eagle and red kite projects: it's a purist point of view, which tends to be forgotten when the purist is watching capercaillies in Scotland, for example. These, of course, were reintroduced from Scandinavia, albeit long ago.

The old female returned in 2002 but laid dud eggs and, after that autumn, was never seen again. The male, though, went on to father a dynasty, and the whole exciting project is written up in the marvellous book, *The Rutland Water Ospreys*, by Tim Mackrill, Tim Appleton, Helen McIntyre and John Wright.

By the end of 2001, 64 young ospreys had been translocated from Scotland to Rutland Water and 59 of them had migrated. No more were collected from 2002, but by 2005 we worried that the majority of the birds which had returned to Rutland Water were males. The American experience had already shown that males are the most likely to return to the area in which they were released. This natal selection is well known in the wild for males, whereas females may turn up anywhere. We hoped that the displaying males at Rutland Water would attract females, and that was what happened in 2003, when the old male attracted a new young, unringed female, which was to stay and breed. More females just called in, had a look, sometimes being fed by males at their nests, and then continued to Scotland, which was really frustrating for the team. From our satellite tracking we also learned that the main track north in spring for Scottish ospreys was through the English/Welsh border lands, so Rutland was a wee bit east to catch the maximum passage. At one of our autumn meetings we decided that we should try to solve this imbalance, and in 2005 I collected 11 young in Scotland and translocated them. Ten of them migrated, but in subsequent years not a single one returned to Rutland Water. One bred at a nest in Scotland and, of course, others could have as well, for not every breeding osprey is checked for colour rings.

Like other projects before, the start of the recovery of breeding ospreys at Rutland Water was a slow process, and although two pairs bred in 2003 and reared five young, there were only single breeding pairs until 2007. From then on, the population rose nearly annually to five pairs rearing 10 young in 2011. And by 2010 six young birds, bred and reared in a nest rather than in a hacking cage, returned to the colony. Tim Mackrill and John Wright documented the increasing numbers and the breeding success, which was often due to their having erected nest poles in secluded locations on farmland with the help and support of local farmers. In 2019 the population reached 10 pairs and reared a total 24 young, while in 2020 10 pairs reared 19 young, bringing the total to 190 young since 2001. Not all of the young returned to Rutland Water, though, and that is another part of the story.

We had refined the techniques so that translocations were successful and could be replicated elsewhere. The key points were to choose the best release sites, build the optimum hacking cages and access fresh fish daily. In the donor sites, it was important to collect only really fit and healthy chicks of at least 320 mm wing length. It was necessary to fit tiny, tail-mounted VHF transmitters, so that fledged young which grounded in leafy bushes or long vegetation in the first days after release could be rescued. We advised against fitting satellite transmitters on translocated young. CCTV installed in the hacking cages proved essential for effective monitoring prior to release.

## Ospreys in Wales

Roger Lovegrove, the former RSPB Regional Officer for Wales, and I had talked on several occasions about the possibility of osprey breeding in Wales, so in early December 1997 I stayed with Roger and his wife Mary at their home near Newtown. It was very cold next morning but with bright sun, and we were away early to meet Dee Doody at a reservoir not far away. I knew Dee from 1989 when he had been our red kite officer in the Highlands. We drove round the reservoir and found three places that I thought were good for building osprey nests, the best on a hillside among stunted larches. Later, Dee built a couple of good osprey nests in trees around the reservoir. We went down into Rhayader for a pub lunch, and called quickly into Gingrin farm to see the famous red kite feeding station, where 20 kites were swirling beautifully over the farm fields, before reaching Gwynffred reservoir. I found one very good Douglas fir tree, suitable for an osprey nest, near the Welsh Water site, and there was another potential nest site in a Scots pine. On 4 December, Roger and I went south to Tregaron – although there, in the marshes, we were actually talking about beavers, a subject we'll be coming to in the next chapter, rather than ospreys. We then visited other potential beaver sites further south, Roger later putting me on the midnight sleeper train at Crewe.

During our travels we discussed the possibility of doing another 'Rutland Water'. Roger was keen, and as he was on the board of the Welsh government conservation body, the Countryside Council for Wales (CCW), he broached the subject the following year with Malcolm Smith, a senior director and a long-time friend, whom we then met at CCW headquarters. In September he wrote to Roger saying he was running into difficulties with the proposal within the organisation. That came as no surprise, as it's often easier to kick contentious suggestions into the long grass. I was back in Wales with Roger in February 1999 to look at areas along the upper reaches of the River Severn, where birding friends of his had seen an osprey between 30 June and 27 August 1998, and the same bird again in 1999 from 6 June to 29 July. It was a female and it had a dark green ring with white letters. I checked with my osprey researcher friend in Germany, Daniel Schmidt, and he reported that it had been ringed as a juvenile on 29 June 1996 at a nest near Lake Müritz in Mecklenburg. That meant that it had arrived in Wales as a two-year-old and returned as a three-year-old, but after meeting no resident ospreys was unlikely to return in 2000.

The 'ospreys' return' idea never went away for either Roger or me, and in 2003 it bubbled to the surface again, as the Forestry Commission were interested in looking in more detail at the proposal. I think my old friend Hugh Insley up in Inverness, the Forestry Commission chief in Northern Scotland and one of the local osprey ringers, had been encouraging his Welsh forestry colleagues. I stayed with Roger again and on 2 October we drove across to Aberystwyth to the Forestry Commission office where I met James Laing, who was charged with the project, along with Tim Blackstock and Catherine Gray from CCW. It was a positive meeting and James's boss Bob Farmer called in at lunch and was supportive. In the afternoon we checked out Forestry Commission woods, finding good nesting areas and fish-rich estuaries, but no perfect hacking site, and also checked the Dovey estuary and Afon Mawddach. The next day we drove to the Betws-y-Coed forestry office and Jonathan Taylor gave us a great tour of the North Wales forests. Once more, there were good nesting sites and at least one good hacking site in Llanrwst Forest, where we hiked through the woods and discussed which trees needed felling to make it perfect for releasing young ospreys. It was again clear that the estuaries were ideal fishing areas, and James Laing asked us to prepare a feasibility document.

I sent the document to James Laing on 11 January 2004, and on 19 March Roger and I met and spent a day answering queries. One of our problems was that CCW said the osprey was not a Welsh breeding species, and it was hard to refute, given a lack of nesting references in the authoritative book, *Birds in Wales*, published in 1994 and written by Roger Lovegrove, Graham Williams and Iolo Williams. My earlier research showed that the osprey had been exterminated in most of the British Isles long before formal ornithological records. James Laing had a bright idea and asked two Welsh-speaking historians to search osprey history in Wales. Caroline Earwood and Tym Elias found many Welsh and dialect names for the species in the literature dating back to 1604, and the richness of names suggested that the peoples of that time knew the osprey. It was not until 1973 that the Welsh name for the osprey, *gwalch y pysgod*, was accepted as the standard. The osprey even appears on the ancient coat of arms of Swansea, apparently dating back to 1316. The osprey had clearly once been a Welsh breeding bird, before it was exterminated. The detailed report of 30 pages by James, Roger and me was submitted to CCW and RSPB in April 2004. We were unaware that the ospreys were already nesting.

The birds themselves had clearly ignored the negativity around their return, for in the summer of 2004 two pairs of ospreys nested in Wales. In June, Steve Watson, a local resident, found a pair nesting in a tall conifer at Glaslyn, near Porthmadog. The RSPB mounted a watch but, sadly, the nest was damaged in high winds and the two chicks fell to the ground and died. Tim Mackrill, John Wright and Barrie Galpin from the Rutland Water osprey team drove over in early July to meet Steve, and with their telescopes were able to read the number on the male bird. It was orange/black 11, which I had collected near Aviemore and translocated to Rutland Water, where it was reared and released in 1998. The female was unringed. Another pair was then located inland in Montgomeryshire, not far from where the German female had summered. They were successful, and on 26 July I met Graham Williams and Blaydon Holt at the Montgomery Wildlife Trust offices in Welshpool, soon joined by Tony Cross, a Welsh ornithologist and bird ringer, and Clive Faulkner of the Trust. We drove to the site, Tony towing a cherry picker to allow us to reach the nest in its dead tree. The female was present when we arrived – I could see a red ring – and then the male flew in with fish. Tony and I went to the nest but, frustratingly, the cherry picker couldn't reach the nest because of a small hedge. Tony could see, though, that there was just one young. I searched below and found that there was an earlier nest, from previous years, in a live Douglas fir close to the dead fir which held the new nest. This meant that the 2004 nest was not the first nesting. When we walked away, the two ospreys settled in the trees and we could see that the male was colour ringed white/black 07 and was a Rutland Water young from 1997, which I had collected in Easter Ross. The female's colour ring was red/white 6J, and she had been ringed in Perthshire in 1999. The male, then, was seven years old and the

The first successful Welsh nest in Montgomeryshire, July 2004. Tony Cross looking into the nest, with one young osprey, from a cherry picker.

female five years old; I was sure this wasn't their first breeding attempt in Wales. White/black 07, if he had come back north as a two-year-old to that valley, might have met the German female, and that would have been another story. The early years of projects often involve chance.

We were absolutely fascinated, and proud, that the Welsh recolonisation was due to birds which had been released at Rutland Water. They had returned to the same latitude but were 200 kilometres west by longitude, and they had been able to persuade Scottish females to stop off on their migration north. As I explained previously, the main thrust of spring migration for returning Scottish ospreys is between Birmingham and the Welsh estuaries.

On 7 December 2004 Roger and I were back in Aberystwyth meeting James Laing at the Forestry Commission offices. After a useful discussion, I went with one of his forestry staff to the Dyfi estuary and found several superb Douglas firs for building osprey nests. The next morning Roger and I drove from his home to Porthmadog to view the osprey nesting area there, and we walked across the fields, with the RSPB man and Steve Watson, to the previous summer's nest tree. It was clear how the nest had fallen out of the 20-metre fir: what a shame it had not been stabilised. I recommended that they should get a climber to construct a nest on the first set of side branches, about a metre or so below the top. Steve said he would organise a tree surgeon to do it before spring. Before leaving, we made a good search by car of the estuary at Porthmadog and I could see many places where nests could be built. My advice was to build a couple of nests, as well as repair the old one, so that if extra birds arrived in spring 2005 they might be encouraged to stay and breed. The nests should not be far from the first one, for ospreys like to be semi-colonial.

Some birders don't like us building nests: they say we should leave ospreys to build their own. We do it because in the natural, non-persecuted, osprey breeding range, osprey nests are used for decades or even centuries, so young new breeders generally find a place to breed on well-established nests, which are much safer, when an adult of their sex has not returned from migration. In fact, it is rare for young ospreys to build new nests, and newly constructed nests are more likely to blow out in strong winds, as happened in 2004; that's why our built nests replicate ancient times and ensure higher breeding success.

Roger and I then drove across country to Welshpool and met Clive Faulkner of the Montgomery Wildlife Trust. We went to the old eyrie there and decided that the dead tree and the nest were secure. If it did blow down in a storm, then the live Douglas fir next to it would make a suitable alternative nest site.

Finally James Laing's feasibility document had been reviewed by CCW and RSPB, and a crunch meeting was to be held on 27 February 2005 at Bangor in North Wales. I asked Ian Newton if he would like to come to the meeting with me, as I was in Birmingham and could collect him from a train and drive across together. It would be a great chance to catch up on all the news. When we got to the hotel we met Roger Lovegrove, and that evening over dinner we discussed many topics and projects. The next morning, we had breakfast looking over the water towards Great Orme's Head and Anglesey, a site of which my diary notes: 'Lots of waders and ducks. Good for sea eagles.' We assembled at an environment building and had a pre-meeting with James Laing and Trefor Owen, which didn't go as well as we had hoped; they had obviously been got at by the Welsh director of the RSPB. At the main meeting we joined David Parker, John Taylor and Catherine Gray of CCW and Reg Thorpe of RSPB. It was a hard meeting, in fact at times downright negative. In contrast with their earlier line – that ospreys had never nested in Wales so should not be introduced – we were now told that they had arrived naturally and that we therefore didn't need to do anything. Our response to this was that it was a tiny population, pioneered by Rutland Water-translocated young, trying to establish itself, and that stochastically they could die out and fail. We suggested that the aim should be a viable population for the long term, not just one or two pairs for public interest at visitor facilities, with the accompanying publicity and membership opportunities. Our aim was to bolster the first pioneers with a translocation of Scottish young over a five-year period. We did finally agree on a translocation and that staff should work together over the preferred release area at Mawddach, but there was no eventual conclusion. We had a pleasant lunch with the group, but I felt that some organisations now lacked enthusiasm; I guessed then that the great days for nature conservation in our country were waning. Ian and I talked on our way back to Birmingham, and our assessment was that nothing more would happen, and that turned out to be true. Another project in the long grass.

That spring, the north pair returned and bred successfully in their reconstructed eyrie and many hundreds of people came to watch them from a roadside visitor watch point manned by the RSPB and volunteers. This pair continued to breed successfully from 2005 to 2010, being the only known pair in Wales, and reared 12 young. No ospreys returned to the Montgomery site, though of course they may have nested in an unknown location. Wales, then, was back to one pair, and if one of the Glaslyn ospreys had died in that period, the Welsh population would have been back to square one. Steve Watson in the north became an experienced osprey nest builder and erected nearly a dozen nests, with some successful breeding. Although the 'official' view was against nest building, an eyrie was erected at the Dyfi estuary on a Montgomery Wildlife Trust reserve, and our good friend Emyr Evans was appointed as the Project Manager. The first pair bred there in 2011, when the breeding female was a Rutland Water bird which had fledged at site B in 2008, so the Rutland reintroduction was still compensating for the failure to allow the Forestry Commission to lead a Welsh reintroduction.

On 18 July I was privileged to see the Dyfi nest with its three young, and met Emyr and Janine. Emyr and BBC *Autumnwatch* had funded three satellite transmitters and had asked me to fit them. The nest is close to the railway line, so next morning we were kitted up in orange suits, safety helmets and glasses to cross the track. Tony Cross climbed the tree, brought down the young, and he and I ringed, measured, weighed and fitted the transmitters. Those three young were tracked and followed to West Africa. To learn more, read Emyr's fantastic illustrated book *Ospreys in Wales – the first ten years*. When there, I was also impressed by the quality of the high-definition CCTV cameras installed by the nest; people saw behaviour close up, especially in the nest, behaviour that I had never seen with binoculars or telescopes. Slowly the Welsh population increased to four pairs in 2014, although the exact number is often unknown; in 2020 five pairs were monitored and reared a total of 12 young. Interestingly, five young from Wales have gone north to breed in Galloway, Perthshire, Cumbria and two in Kielder Forest. Thousands of people now enjoy watching ospreys in Wales, sometimes quietly as they fish at estuaries and lakes, or from the impressive new osprey visitor centre overlooking the Dyfi estuary and its osprey nest. I wish we had released 60 young in Wales at the beginning of the century: the situation would be different now, and it's important to remember that the osprey is still a rare breeding species in England and Wales. To this day, there are huge areas of the country still lacking the fishhawk, an icon of our spring and summer skies.

## Red squirrel recovery

In the early 1960s I became used to the sight of George Waterston arriving at the Loch Garten osprey reserve, where I was RSPB warden, with a creature for me to care for. One day, he pulled up in his Dormobile and fetched out a small box, in which he had a baby red squirrel. While driving north, he had seen a dead adult squirrel on the road near Pitlochry and, next to it, just the glimpse of a movement. Stopping to walk back, he had found that a car had killed the mother – presumably as she moved her young between dreys, as squirrels often do – but that the baby had survived and remained beside her. I soon had the young squirrel installed in a comfortable little hutch and fed it milk with the rubber filler from an old fountain pen. The squirrel, whom I named Timmy, thrived. In no time at all he was a favourite at the osprey camp and he came with me on osprey watches, made appearances from within my pullover at the local pub, jumping from person to person, and even came with me high in the Cairngorms. As he got older he would venture away from the caravan into the local Scots pine trees and, finally, we lost contact. I hoped he was with the local squirrels, of which there were many.

When I returned from the Shetland Islands to work in the Scottish Highlands, I would commonly observe red squirrels when I was carrying out bird fieldwork and would occasionally find them dead, as prey, in golden eagle nests. They were most common in the big areas of Scots pine forests but were lacking in some outlying areas. Their history in Scotland is intriguing, for they were common through the first half of the century, having been all but extinct in the early 1800s, with just possibly a small remnant population in the old firs of Abernethy Forest. Their fortunes had changed dramatically in the late 1880s, when Lady Lovat of Beauly pioneered the translocation and release of English red squirrels. Other landowners joined her to spread squirrels to the most suitable woodlands, and the squirrels responded by breeding rapidly and spreading across the countryside. The numbers became so high that special squirrel clubs were established to kill them, so that the new woodlands were not devastated.

An examination of the records suggests that red squirrels disappeared from various outlying locations in the latter half of the 20th century. There is no scientific evidence of the process of this local extinction. It could be the result of chance, because of the low population sizes involved, and it was almost certainly a result of the very low level of woodland cover following the increased levels of felling – especially of pinewoods – which took place in the last war. I also suggested that there was the possibility that a resurgence of pine martens had an impact. Until the 1970s, this species, which is a predator of red squirrels, was extremely rare and mainly confined, because of human persecution, to remnants in the Western Highlands. With a rapid increase in forest planting from the 1960s, and a changed attitude towards the protection of martens, the species flourished and rapidly expanded its range east and south. By the 1970s and '80s, red squirrels had not only declined to tiny, remnant populations, but disappeared in outlying woodlands in the northwest Highlands. Many of these squirrels may have had no previous experience with martens and, as a consequence, would have been easy prey. Once the martens reached areas of much larger populations of red squirrels, there was a chance for the two species to once again reach a natural balance, because of course they had both evolved in woodland ecosystems.

In 1999 I received a request from a landowner, Hugh Tollemache, who wished to have red squirrels in woodlands on his property at Ben Shieldaig. I replied, saying that I was in sympathy with his wishes and thought it was feasible. I was prepared to help, and it was probably easy to do, but the project would require field surveys, to check that there was sufficient habitat and food, as well as various approvals and licences, and I knew that it might entail a lot of bureaucracy. In July the landowner abandoned the idea because his forestry consultant thought the area too small. He also added, 'The legal aspects also appear awesome to overcome.' How right he proved to be.

In late October 2000, when visiting the Corrour estate, Ted Piggott, the head stalker, told me that the owners would love to see red squirrels in the conifer woods beside Loch Ossian. This is a remote area in the Lochaber mountains, east of Ben Nevis, and known, if it's known at all, because of the remote Corrour train station on the Fort William line. I had regularly carried out fieldwork there, monitoring black-throated divers as well as the local peregrine and golden eagle eyries. The landowners at that time, Donald and Margo Maxwell MacDonald, were keen on wildlife. I agreed to help and the following year surveyed the woods around the loch to ascertain their suitability for squirrels before starting to prepare a feasibility plan.

In 2002 I submitted to Scottish Natural Heritage in Fort William a proposal to establish a population of red squirrels at the Corrour estate. Corrour is famous for being the site of one of the earliest attempts at re-forestation of degraded lands in the Scottish Highlands. Sir John Stirling Maxwell, owner of Corrour from 1891 and ancestor of the then owners, was a champion of afforestation in Scotland, and from 1892 planting was started around Loch Ossian. Seventy-one different conifer species were planted, as well as beech, oak, sycamore and other broadleaves. Because of this early pioneering work, the woodlands at Corrour were highly diverse in species, age, height and structure. There were approximately 320 hectares of woodland extending 5.5 kilometres along the northwest shores of Loch Ossian to an average width of 600 metres, and 4.5 kilometres along the southeast shores to an average width of 400 metres. About 50 per cent of the woods were pre-1950s planting, 20 per cent between 1950 and 1980, and 30 per cent post-1980s planting. Broadleaves accounted for about 15 per cent of the total, pure stands of Scots pine for 15 per cent, but much of the mixed conifers (30 per cent or more) also included Scots pine. The plantings in the late 1980s were mainly of Sitka spruce and lodge-pole pine.

My expert view was that the Corrour woods were suitable for red squirrels. The very diverse woodland habitat was highly appropriate for squirrels, with plentiful sites for building dreys. There was a good to excellent food supply, and the woods also provided a range of alternative foods, including berries and fungi. I also noted that the very large, high and multi-branched conifers would provide high-quality, above-ground habitat for red squirrels and that their relatively dense structure would provide greater protection from predators, such as pine martens and common buzzards, than in open, single-species Scots pine or larch woods.

I recognised the isolated nature of the location and that it was not possible, then, for red squirrels to get there naturally; but I considered it a good location for a translocation. The head stalker and his staff would do everything possible to ensure success, including the provision of supplementary feeding – nuts and seeds – during the acclimatisation period and subsequently, if required. There were many potential donor sites for catching animals, including in Strathspey. The land agents sent me accurate data on tree species and extent, from which I estimated that the Corrour woods could be home to around 256 full-grown red squirrels, with variations ranging from 64 to 512 animals. These numbers would maintain a viable population. This population would have two additional future advantages for red squirrel conservation; its isolation could provide a valuable refuge, which would be safe from incursions of grey squirrels, and it had an isolation benefit during squirrel epidemics, especially that of the grey squirrel parapox virus.

In Scotland – then as now – the law protected red squirrels, so a licence from Scottish Natural Heritage was crucial. The species was listed on Appendix III of the Bern Convention and was protected by Schedules 5 and 6 of the UK Wildlife and Countryside Act. It could not be intentionally trapped, killed or kept, and the dreys could not be disturbed except under licence from SNH. It was also a UK Biodiversity Action Plan priority species. They were also protected under the Wild Mammals (Protection) Act, 1996. In 2004 red squirrels were to receive additional protection from the Nature Conservation (Scotland) Act, 2004. Subject to certain exceptions, it was an offence to intentionally or recklessly kill, injure or take a red squirrel; to damage, destroy, or obstruct access to any structure or place which a red squirrel uses for shelter or protection; or to disturb the red squirrel while it is occupying a structure or place which it uses for that purpose.

I sent an initial feasibility document to Ted Piggott on 23 July 2002, with a recommendation to the owners that, before working up a detailed programme of practical activities and a timetable, I should discuss the proposal with Scottish Natural Heritage in Fort William to gain their support and, ultimately, the award of the necessary licence. This meeting occurred on 26 August 2002, when I discussed the proposal with Greg Mudge and Christine Welsh; we had a good discussion that I thought was reasonably positive. Later in the autumn, after they had consulted their mammal research staff in Edinburgh, we received the very disappointing news that our proposal had been turned down.

Two years later, the species came back in the frame, when I was asked about the possibility of a red squirrel translocation to the Dundonnell estate in Wester Ross. On 19 May 2004 I visited this beautiful part of the west coast, where Jane Rice, the wife of the owner, was very keen to reinstate red squirrels, which had disappeared a few decades before. I had a very interesting tour of the woodlands, along with Steve Potter, and by the time the three of us stopped for tea, before I headed home, I was convinced that it was a viable proposition. Nevertheless, after the disappointment in Lochaber, I told Jane that, although it was possible ecologically, it might be difficult to get official authority to start. I slowly began to assemble my ideas and was ready to take it forward in 2005, receiving encouragement from the owners to proceed during a full-day field visit in mid-summer, when I visited all the wooded areas and took notes and a series of photographs.

Dundonnell is famous for its fine house – dating back to 1769 – and gardens, set within a designed landscape. The walled garden was established in the early 19th century, when woodland planting of introduced and native species was also carried out, while an arboretum established by Alan Roger, one of the previous owners, is a more recent addition. As a consequence, the variety of tree and shrub species is extensive. Part of the location was designated as the Dundonnell Woods SSSI for gorge and hillside woodland, and contains perhaps the best example of western gorge woodland. The hillside included both ash and alder. In modern times, various plantations of mixed conifers had been planted in the main valley, and the interconnected woods on Dundonnell, which extended for about 7 kilometres, included native and planted broad-leaved woods, as well as planted Scots pine and other conifers of various ages. There were at least 400 hectares of woodland within the estate and additional woodlands of approximately 200 hectares of mixed conifers in neighbouring land, including 50 hectares of mature, planted Scots pine. Semi-natural woodland (birch, ash, sessile oak, hazel, rowan, aspen, wych elm, willows and other species) accounted for 250 hectares. Conifer plantations established between 1957 and 1962 (so by then 40–50 years old) were made up of 15 ha Scots pine, 35 ha lodgepole pine, 26.5 ha Sitka spruce, 20 ha of European, Japanese and hybrid larch, plus small amounts of other conifers.

The policy woodlands – those associated with the house – were up to 200 years old and featured, in particular, beech and European larch, but also included oak, ash, elm, lime, sweet chestnut, horse chestnut and southern beech. Most of the woodland was old, but a quarter was planted between 1957 and 1962. There was an active forest plan in place, encouraging a more natural and diverse woodland and, outside the policies, favouring native species. On field surveys I noted that the northern plantations were very mixed and that seed crops were present in lodgepole pine, Sitka spruce, Scots pine and larch. I also noted that larch was seeding well naturally.

The western woods near the roadside at Darach included 100-year-plus Scots pine, as well as more recent lodgepole pine,

Scots pine and larch, with birch and rowan. The large Scots pine-wood on the west side of the road had trees dating back to the same era, with a good cone crop in 2005, and also evidence of the 2004 crop. There was much mature alder and other native broad-leaved species through the gorges; and the plantations furthest to the south, opposite the high cliffs, were a mix of 30-year-plus larch, lodgepole pine and Sitka spruce.

I concluded, in my feasibility report, that Dundonnell had a very diverse woodland habitat highly suitable for red squirrels, with plentiful sites for building dreys. Many of the woodlands had a warm, south-facing aspect and provided a good living environment. The whole woodland was contiguous, approximately 6 kilometres north to south, allowing squirrels to range from one end to the other. The wide variety of tree species enhanced the opportunities for seed production as an excellent food supply, especially in years when one species may fail; it would also extend the season of cone and seed availability. As I had found in that original plan for Corrour, years before, the woods also provided a range of alternative foods, including berries and fungi (red squirrels often cache fungi in trees for use throughout the year). Finally, the very large, high and multi-branched conifers and deciduous trees, as well as the dense plantations, provided high-quality, above-ground habitat for red squirrels; and their relatively dense structure would provide greater protection from predators, such as pine marten and common buzzard, than would open, single-species Scots pine or larch wood.

As I examined the status of red squirrels in the Highlands, it was clear that the population was still eroding around the outer edges of the recent distribution, such as in southeast Sutherland, despite the increase of seed-bearing conifers since the 1970s. I considered that there was an important piece of restoration to be done, but that it was greater than just Dundonnell: our project there should be a precursor of an important addition to red squirrel conservation in Scotland. My report widened the scope by stating, 'The vision would be to re-establish red squirrels in suitable forests and woodlands to the north and west of the present range, in order to increase the distribution and overall population and to create refuges free from grey squirrels and associated diseases.' The UK Action Plan for red squirrels listed as the third of its objectives 'to re-establish red squirrel populations, where appropriate'. Our project presented the case for translocating red squirrels to Dundonnell in order to restore the species to that area, and to gain experience in translocating red squirrels into areas free of both grey and red squirrels. The need for such knowledge and experience is also referred to in the UK Action Plan. What we were planning to do, then, had national implications.

The drastic decline of red squirrels in the UK, to the point at which it was now considered vulnerable, had been caused partly by the fragmentation of suitable habitats, a lack of food, disease and, in some areas, road casualties; but the principal threat was continuing to be competition from the alien American grey squirrel, and that threat was growing. First introduced to Britain in 1876 and now numbering over two million, grey squirrels had been common south of the Highlands, but they were now, sadly, spreading north from Perthshire, and outwards into the watersheds of the Don and Dee rivers from a population in the city of Aberdeen. By 2006 the situation was so grave that a whole range of new projects and proposals was initiated, including Biodiversity Action Plans, Red Squirrel Groups, conferences and government intervention. In February of that year, at an international conference on red squirrels, the Deputy Environment Minister for Scotland, Rhona Brankin, gave the stark warning that 'Scotland's red squirrel population is likely to be extinct by the end of the century unless urgent action is taken to conserve the threatened native species.' She also said: 'I am proud that Scotland is home to 75 per cent of the UK's red squirrel population and, in my view, our efforts here today represent the last hope for this iconic species.'

Despite that ethos, the provision to Scottish Natural Heritage of my comprehensive feasibility report outlining how we would carry out the translocation to restore red squirrels to the Dundonnell woodlands was again proving contentious. We were finding it very difficult to get a clear response from SNH, yet we required a licence to live-trap red squirrels, so their support was essential. By this time, I was getting frustrated because I could not see why we couldn't move ahead. So, on 1 March 2007 – nearly five years after our first discussions in Fort William – it was agreed that we would meet at their HQ, Great Glen House in Inverness. The regional manager, Lesley Cranna, and Tamara Lawton from the Ullapool office were present, and we had a video link to Dr Mairi Cole, a mammal scientist at SNH in Edinburgh. Tamara had helped clarify matters during the writing of our report and Lesley had been very helpful in explaining the pros and cons. Nevertheless, I was still not getting definitive feedback from Edinburgh. At one point in the meeting, I became so disheartened that I said I didn't want to continue the discussion but that, instead, I would leave immediately and discuss the project with their Chairman. This appeared to break the logjam, for our discussions noticeably improved and, by the time I left, I felt – at last – that we were on the home run.

I again updated our feasibility document and when I next met Jane Rice, on 13 June, I was able to tell her that it looked like we were able to start. That day I carried out fieldwork on seed availability in the different tree species and on the presence of pine martens. I also talked with Johnny MacSporran, the retired keeper, who was able to give information on the loss of squirrels in the 1970s. On 18 September 2007 I outlined the proposal to a meeting of the Highland Red Squirrel Group and we had a very useful discussion, after which I produced a final version of the proposal. I was encouraged to sense that there was real interest and support from the members, as well as a clear recognition that there were considerable areas of suitable woodland presently unoccupied by red squirrels, and that this project offered an exciting way forward. Half the members present offered help with practical aspects, if and when the project got underway, and the group wished to be involved. At that time, the Highland Red Squirrel Group was a partnership of members of the public, Forestry Commission Scotland, Highland Council, Scottish Natural Heritage, Scottish Wildlife Trust and the Highland Biodiversity Partnership. The Chairman was Ian Collier, based at the Forestry Commission office in Dingwall, along with Juliet Robinson, the Red Squirrel Conservation Officer. An Action Plan was produced by the group which aimed to deliver the species national action plan across Highland Region, the last significant 'grey-free' refuge of the red squirrel. This unique status allowed effort to concentrate on consolidating and improving habitat for red squirrels rather than having to fight a rearguard action against the spread of greys. Noticeably, translocation was not included in the actions.

Everyone at the Dundonnell estate was determined to do everything possible to ensure the success of the project. Its aim was to establish a viable population of red squirrels in the wild, with minimal long-term management. We considered this highly achievable in this location, where there was to be no competition from grey or red squirrels, with no important species prejudiced by their return. The very isolated nature of

Dundonnell, though, would mean that any future red squirrel population would also be isolated, which would bring disadvantages as well as benefits. The benefits would include the freedom from incursions of grey squirrels and diseases related to squirrels, while the disadvantages sprang from a lack of connectivity, especially for the renewal of numbers after natural lows. This, though, we viewed as a minor problem, which could be addressed by a degree of wildlife management. We estimated that Dundonnell could be home to 320 full-grown red squirrels, with variations ranging from 80 to 640 animals. These numbers would easily maintain a viable population, and the increasing amount of woodland available in the future would enhance these population estimates.

The estate was prepared to provide nuts and seeds to the squirrels as supplementary feeding if they needed it, but first we had to decide where the squirrels would come from. The donor sites would be discussed with SNH, the Highland Red Squirrel Group and other organisations involved with squirrel conservation. There were many possibilities in Badenoch and Strathspey, Nairn, east Inverness and Moray, where I lived: all areas free of squirrel pox and where the animal was common. Squirrel numbers are highest in mid- to late summer and into the autumn, before winter losses make their mark, and as many squirrels come to garden and visitor-centre feeders, it would be easy to take animals without impinging on designated nature conservation sites. This would be especially appropriate at the end of the visitor season, when some of these facilities stop feeding. It might also be advantageous to trap squirrels in forests prior to active felling programmes. It is well known that red squirrel populations rise and fall in cycles in response to the availability of their main food, conifer seeds, which has cycles of abundance. The evidence shows that this rodent species has the ability to bounce back in years of plentiful food supplies; in consequence, any small numbers lost locally to translocation would soon be made up, with a negligible effect on local populations.

In the early stages of the translocation project released squirrels would be monitored. The use of implanted pit-tags, as used in cats and dogs but much smaller, would be possible, with a few animals radio-tracked. DNA and blood samples would be taken by a veterinarian before release, and drey counts would be carried out in early winter.

Historic reintroductions, such as that carried out by Lady Lovat, had clearly been successful, but a more recent translocation of 44 red squirrels from Fife to West Lothian in 1994–5 had resulted in seven deaths in transit through stress. This was raised as a problem, but I considered that we could translocate much more successfully. I contacted my friend Derek Gow, an expert on captive mammals, and he recommended that the carrying box for each squirrel should be big enough to hold a football, and that cut-up, juicy apples as well nuts should be available during transit. A larger project had been carried out at Thetford Chase in Norfolk in 1993–6, but the major problems there had been associated with the presence of grey squirrels.

Our feasibility report covered all the matters raised in the IUCN guidelines, which encourage best practice in reintroduction and translocation projects and weed out unsuitable proposals. The owners confirmed that the release area and future red squirrel population would be assured of long-term protection, and we found that there was great local support in restoring red squirrels: I was also very encouraged that so many people who fed squirrels in their gardens were prepared to allow the live trapping of donors. There would be a presumption that if red squirrels from this population were required for future translocation projects, donor animals would be available.

At Dundonnell the owners and I identified two ideal sites within the forest for building large 'aviaries', to hold translocated squirrels for a 'soft' release. Red squirrels trapped at feeders in agreed locations in Badenoch and Strathspey, Moray, Nairn and Inverness would be housed in these very large cages until they could be 'soft released' – kept initially in captivity for a short time before being let go. Our initial thoughts were to move 30 to 40 red squirrels, and to release the majority by soft release at the two sites, but to release the rest directly into the forests at locations at the other end of the woodland. Squirrel feeders would be located at various sites throughout the forest and kept full of nuts. Slowly, over time, the provision of nuts would be reduced, especially to coincide with heavy cone crops. This was also a time to learn from those who had recently worked with red squirrels.

On 13 September 2009 I caught the early morning plane to Manchester and picked up a hire car to drive to Anglesey. I had arranged to visit Dr Craig Shuttleworth, who had carried out his PhD research on Lancashire red squirrels and was running the Anglesey red squirrel project in Wales. During our day tour of this beautiful island, Craig gave me a brilliant chance to catch up on squirrel knowledge and field experience. We talked through all aspects of our project; he recommended buying squirrel live-traps, usually used for greys, and – once the squirrels had been caught – keeping them calm by covering the trap with dark cloth. The animals can be moved from the trap into a large canvas sack, and then transferred into an inspection tube, 23 cm long and 5 cm in diameter, made of weldmesh. This is a safe way to inspect, weigh and measure squirrels without harming them. The squirrels can then be run into plywood nest boxes, one for each squirrel, with a 6 cm hole in the top corner. The nest boxes should be filled loosely with dead grass and moss and, in transit, contain cut-up apple and nuts.

Once at the release area, the boxes should be fixed to trees, the exit hole exposed but with a little moss poked in, which the squirrel can push aside when the coast is clear. The feeding boxes would be standard Perspex-fronted ones, filled with peanuts. Craig described the soft release cages and recommended that the squirrels stay there for no longer than two weeks; he also showed me how to use pit-tags for identification as well as radio collars. The day was extremely useful and Craig offered to give advice at any time. I recognised that he had a deep and practical knowledge of red squirrels, and that he was a very important ally for us in this work.

Autumn 2009 saw the project underway. I visited Dundonnell on 28 September and had a very useful planning meeting with Jane Rice and her son, Donald. We visited the excellent soft release cages built at the edge of two woods; the smaller contained two squirrels which had been orphaned and looked after by the Highland Wildlife Hospital in Ullapool. They would be released before the translocated squirrels arrived. The estate keeper, Alasdair Macdonald, who proved to be a great asset to the project, would top up the peanut feeders and keep a record of sightings of squirrels throughout the first winter. I checked seed availability and found that both Scots pine and larch had good cone crops.

On 21 October I spent the day at Dundonnell with Alasdair, making certain that everything was ready for the arrival of the squirrels. Richard Page at Dundonnell had built a good supply of squirrel feeders and squirrel nest boxes, similar in design to those used on Anglesey; I had purchased 20 squirrel live-traps, and built a good supply of squirrel feeders and large nest boxes. On the way home I detoured to Ullapool for a discussion with

Tamara Lawton, who told me that everything was agreed and that SNH had prepared a press release.

I decided to catch a small number of red squirrels to test our procedures. The village of Boat of Garten was chosen for trapping, and in mid-October I visited Allan and Heather Bantick to discuss suitable catching sites and to give them six traps. These were pre-baited, in non-catching mode, so that the squirrels would get used to them. On 25 October, at first light, the traps were set for real. During the morning, Allan and Heather, with the help of friends, caught four squirrels in and around the village, I caught one squirrel near our home at Dunphail, and another next morning. The squirrels were kept individually in

Red squirrel in live trap in Moray, prior to translocation to Wester Ross, 2009.

nest boxes which had been lined with fresh hay, and supplied with a variety of nuts, sweet apple and carrot.

Early next morning I drove over the high road past Braemore Junction and down towards the woodlands below An Teallach, the stark mountain streaked with overnight snowfalls. The last twists of the road took me down to sea level and the soft release area. With Alasdair Macdonald, I placed two nest boxes in each section of the large cage, and the other two in the small cage, from which the orphan squirrels had already been released. Squirrels caught at the same location were kept together, and a plentiful supply of food was provided in the feeders. These squirrels were soft released 11 days later, before the arrival of the next group.

The next catch was after dawn on 5 November, with trapping sites organised in Carrbridge, Cromdale, Grantown-on-Spey and Dunphail by Frank Law, Bill Cuthbert, Stephen Corcoran and myself. Jane Harley, a veterinarian, and Gaby Bongard, a veterinary assistant, of Strathspey Veterinary Centre in Grantown-on-Spey, had very kindly offered facilities to check the health of the squirrels. They were joined by Anna Meredith of the Royal Zoological Society of Edinburgh and Edinburgh Veterinary School, who had organised a protocol whereby each squirrel was sexed, aged, measured, weighed, fitted with pit-tags and recorded; blood and hair samples were taken for analysis, and notes were made of their condition, parasites and any features. Each of the 18 squirrels was then placed in a uniquely numbered nest box so that individuals from the same catching sites – a maximum of two from any trap site – could be kept together at release. Three squirrels had been fitted with BioTrack radio collars.

Bill Cuthbert and I drove to Dundonnell on 6 November and met Jane Rice and Alasdair. Six squirrels in their individual nest boxes were placed in the now-empty release cages. Four squirrels were located in their nest boxes at a release site to the east of the top field. Four nest boxes, containing squirrels, were placed, in a loose group, high in trees at an ideal location with Scots pine, larch and Sitka spruce. The other four squirrels were located in their boxes around a clearing in the pinewood to the east of the small release cage. This was another ideal site, with a mix of Scots pine, larch, Sitka and Norway spruce and large oak and birch trees. Six or seven squirrel feeders were placed at each release site and filled with a mixture of nuts and maize. Alasdair reported that the squirrels released earlier were regularly feeding at feeders near the small release cage.

The veterinarians were unable to give another day to the project until 21 November, so un-set traps were put out on the previous days at Logie, Edinkillie, Grantown-on-Spey, Cromdale and Carrbridge. That morning 14 squirrels were trapped and taken to the Strathspey Veterinary Centre for examination and marking. All squirrels were again chipped with Trovan pit-tags and housed individually overnight in our shed, in nest boxes supplied with nuts and fresh apples. Once again, all the squirrels were in excellent condition and the following morning, when I arrived in Dundonnell to meet Alasdair, we found that all of them had travelled well and were in great form.

Six squirrels were placed in the soft release cages, the squirrels from the 5 November translocation having been released. Additionally, eight squirrels, in two batches of four, were placed in a wood to the west of the sawmill. The lower four nest boxes were placed in trees at the edge of the forest of larch and Sitka spruce, while the other four boxes were placed further up the hill in the Scots pinewood. As previously, squirrel feeders were placed in trees around each release site. The last two squirrels were placed in Dundonnell House gardens, where at least one red squirrel had taken up residence and was feeding at the bird feeders. One nest box was placed carefully in the famous 2,000-year-old yew, the other in a large holly; additional squirrel feeders were also put up. All of the squirrels, except the first six, were fitted with small animal pit-tags for future identification, while three were also fitted with radio-tracking collars, with one more to do. Anna Meredith would be researching the blood samples, and the hair DNA samples would be retained for future studies. It was very encouraging to see that all the squirrels were in excellent condition, with just one having a high infestation of lice, which was treated.

Since the start of the project, Alasdair had maintained the feeders and checked for usage, recording all data gathered during these visits, including sightings (date and place) of squirrels. On 1 December, one was found dead in a nest box in the larger cage and a subsequent post-mortem suggested that it had died of stress, possibly through conflict with the other squirrel. After this, I decided that we would do no further soft releases (keeping in cages to acclimatise): our experience with hard release was very good, and it was much easier and less stressful for the animals.

In view of the lateness of the season, I decided to delay catching the remaining 18 squirrels until the days started to lengthen in February. We held a licence for 50 animals and had proved that we could catch and translocate the squirrels in really excellent health and condition, without losing any of them due to trapping, examination and translocation. The mean weight of males was 334 grams, and of females 318 grams, which was above the recorded mean.

We got a real surprise when one of the red squirrels was seen on the western coast of Loch Broom in the township of Letters, 6–7 kilometres away over the mountains. It visited several gardens, continued around the loch and finally settled in a garden, surrounded by woods, at Leckmelm. The garden belonged to John and Anne Lychet, who put out feeders for birds, and the squirrel liked this easy supply of nuts in a safe, wooded landscape. I visited John and Anne on 18 January and saw the squirrel's favourite bird feeder fixed to a large redwood tree across the lawn. The squirrel survived into the late winter and, knowing that I had some more squirrels to take to Dundonnell in February, I thought of releasing one to keep this lone animal company. I asked John if he could take some photographs to see if I could sex the individual, and the photos clearly showed that it was a male. On 20 March I trapped the final eight squirrels near my home in Moray and in Strathspey, and took them – five males and three females – to the veterinarians, where Jane and Gaby checked them all over and found that one of the females was pregnant.

We set off for the release area in Dundonnell on the afternoon of a bright, sunny day. There, Alasdair and I carried out our normal procedure, placing four boxes in one area and three in another, along with eight nut feeders, and I kept back one female in her box. This completed a release of 43 squirrels at Dundonnell as the starter population.

By the time we reached John and Anne's house it was dark, so I carefully fixed the nest box, with the female in it, on the big redwood in the garden. On this occasion I left the covering over the exit hole and attached to it a long string which ran across the lawn and through the kitchen window. I suggested that, next morning, as soon as they saw the male coming to the nut feeder, they should allow the female to come out by pulling the string. If the male did not arrive, though, they should release the female at midday; of course she had plenty of nuts in the box, as well as cut-up apples, to keep her fed. All morning, there was no sign of the male, but just at midday, as John and Anne were preparing to pull the string, there he was. They quickly opened the nest box, while the male came to the feeder, ran to the nest box and then back again. The female came out and the pair of them ran round the tree, the male chasing the female. The female went back into the box, presumably to get food, and on the following days the pair were seen together on many occasions, and later that summer young squirrels were also seen in the garden trees. They, in turn, bred in subsequent years. Local people were thrilled to hear that squirrels were again breeding in the woods near Ullapool and described our intervention as a kind of 'Lonely Hearts Club' for squirrels.

* * *

Alasdair diligently kept a detailed record of squirrel activity, and our telephone conversations were very encouraging, for the project was going well. On 2 April I was very pleased that Lesley and Tamara from SNH could join us and see progress. We took them round all the sites where squirrels had been released and, by using our VHF receiver, located three of the radio-collared squirrels in the main block of conifers and, later, the remaining female which was visiting the feeders in Alasdair's garden. We noted that she was large and clearly pregnant; everything was looking good. On 18 May I returned to meet Jane and Steve and, once again, we were able to find all four radio-collared squirrels, while transects showed that the squirrels had been busy on Scots pine and larch trees. Meanwhile, Alasdair had reported the first signs of young squirrels. On 18 June we were delighted to host a visit by Ian Jardine, the chief executive of SNH, with his deputy Susan Davies as well as Lesley and Tamara. Alasdair gave us a tour of the woodlands; they saw a couple of squirrels as well as the cages, nest boxes and feeders. While walking the woodlands we discussed the processes, the successes and the difficulties, one of which was how really to work out the population size using the suggested methods of monitoring. I considered these more appropriate to lowland English woods than to conifer woodlands on hillsides covered in heather.

During 2009 the squirrels started to move into new areas and were seen at feeders in several gardens, including the keeper's on the next estate. Young squirrels were seen, and once the leaves fell off the deciduous trees, we started to see the hidden dreys built by the squirrels for having their young. In 2010 squirrels had another successful breeding year and were really obvious to people in the district, although sadly a few were killed by cars, despite Alasdair's warning signs for motorists. The Dundonnell estate organised a biodiversity weekend with SNH in August and we had a very interesting two days, with four evening lectures at the Dundonnell hotel, talks and meetings in a marquee in the garden and guided walks for locals and visitors. Many squirrels were observed, with the best count of 12 seen on a walk led by Juliet Robinson, the Highland Red Squirrel Group coordinator. Jane Rice and her son Donald were excellent hosts: for them it was an ambition realised to have red squirrels back.

By the autumn of 2010 Alasdair and I knew that we had a lot of squirrels ranging through the whole of the woodlands around Dundonnell. We frequently saw them and, wherever we walked in the conifer woods, found evidence of them having gnawed on cones to extract the seeds. We considered that there were hundreds present but could not find a method to ascertain the true population. I read about the suggested methods but could not see how they would work in West Highland conditions. In fact, it was impossible to find examples of successful estimation of population size of squirrels in any Highland forests. When I talked with squirrel experts, there was always uncertainty over how to judge the population size. Was it by the numbers present in March, before the breeding season? Was it by the numbers at the end of the summer, which included all the young? Or was it the numbers in midwinter? What was clear was that red squirrel numbers fluctuate incredibly based on the availability of seed, especially Scots pine. People often forget that the squirrel is a rodent, and that peaks and troughs are part of their lives.

Because the tiny population at Leckmelm was still surviving, I wanted to move some of the Dundonnell squirrels into the next watershed, running from Braemore Junction down through the conifer woods to Ullapool. In March 2011 Alasdair and I decided we would carry out a live-trapping trial, so I went over to stay two nights in the estate cottage. On the 16th we caught 12 squirrels which we measured and weighed, and I did an interview on our project with Dougie Vipond from BBC TV in Aberdeen. The next day brought our total to 18, so we knew that we could catch squirrels, when needed, for future translocations. The squirrels had another good breeding season in 2011 and numbers were even higher; we thought, in fact, that in the whole area there might be as many as 500. The project had clearly been very successful.

On 1 November we arranged a meeting at Dundonnell with Jackie Mackenzie and Juliet Robinson, coordinator of the Highland Red Squirrel Group. Becky Priestley and Fiona Newcombe, who were working with me at the time, also attended. We visited all the woodland areas, examined food supplies, checked dreys and debated future options. These were discussed at a meeting of the Highland Red Squirrel Group in Dingwall the following evening, at which we proposed future translocations. We received a positive response from all the members, in contrast with the reception we'd been given at a meeting earlier in the day with SNH, where the science and degree of monitoring were still bones of contention. On 13 January 2012 the squirrel team carried out a series of transects for signs of gnawed cones throughout the woodlands, and we then moved to the Leckmelm valley, where we identified small numbers of gnawed cones.

Two weeks later I met Ron Macdonald, SNH senior manager, to discuss how to resolve the dispute over the success of the Dundonnell Red Squirrel Translocation Project. My Foundation (RDWF, see page 431 for more information), the Dundonnell estate and the Highland Red Squirrel Group regarded the project as very successful in all three summers, from 2009 to 2011, while SNH were concerned that the success predictions lacked a 'science base'. We maintained that the monitoring of squirrels in the habitats at Dundonnell was extremely difficult and lacked the ability to establish exact numbers, but considered that the reports we had submitted to SNH established the fact that squirrels had increased fivefold or more. In the search for a defined number, Ron agreed that a scientifically proven number of 50 or more would prove that squirrels had increased and survived as a population for three years.

I decided that we would prove this by carrying out two days of live trapping, marking and release, with a repeat two weeks later. Alasdair started to set out traps from 5 February and the live trapping, under licence, was carried out on 8–9 February and 24–25 February by Alasdair, Becky, Fiona and myself. All trapped squirrels were sexed, weighed, assessed for health condition and breeding state, and marked by a unique combination of marks 1 cm wide snipped at nine different positions in the tail fur. They were then immediately released.

The trapping sites only covered part of the whole site, but a total of 95 squirrels were caught, of which 64 were new individuals and 31 were re-traps; 35 females and 29 males were trapped. Two were sub-adults; the rest were adults. All trapped squirrels were in excellent condition with clear, bright eyes, clean ears and excellent skin. Ticks were found on four squirrels. The mean weight of trapped squirrels was 324 grams for males and 324 grams for females, which is well above the mean weights recorded in England – males 279 grams and females 278 grams. The number live-trapped exceeded SNH's requirement and on 19 February I was informed that we could apply for additional licences.

We could now, under licence, trap squirrels to translocate 12 kilometres or so over bare moorlands to the next watershed. Alasdair set out 32 traps in 11 previously-baited sites across the Dundonnell woodlands. Traps were prebaited with a mixture of peanuts and black oil sunflower seed for three days beforehand and secured in the open position with cable ties. On 30 March Becky, Fiona and I set off for Dundonnell and were soon checking the traps. By mid-afternoon we had caught 20 squirrels – 11 males and 9 females – for short-distance translocation. Previously, Alasdair had introduced me to the keepers on three estates in the Loch Broom valley and we had been given permission to release squirrels. The squirrels were released across eight sites on three estates: eleven on Braemore, five on Inverbroom and four on Foich. These translocated squirrels thrived, finally joined up with the Leckmelm population and, within three years, red squirrels were distributed from Braemore to Ullapool. The people of Ullapool were particularly pleased to see red squirrels visiting their gardens and nut feeders. Our second translocation was successfully completed.

* * *

For many years I had visited the Alladale estate, where the owner, Paul Lister, entertained exciting ideas of having wolves and brown bears living in a massive, fenced estate. He had also become interested in my red squirrel translocations and said he would like to have them back in the Alladale pinewoods. On 5 April 2011 I carried out fieldwork to check the suitability of the area. I discussed red squirrels with Hugh Fullerton-Smith, Innes MacNeill and David Clark, who were all enthusiastic, and I checked out the main pinewoods. On my way up the glen I had stopped in several woods in the Langwell Strath to check for gnawed cones, in case squirrels were present, and also to assess squirrel food supplies. Ten years or so previously I had seen a red squirrel in the Rosehall area but was informed that the species had disappeared at the end of the last century; previously squirrels had lived all the way north to the woodlands around Loch Fleet.

My next call was to Amat, a favoured squirrel location in earlier times, to discuss the idea with the owner, Jonny Shaw. He liked it and told me that when he had come home to take over the estate in 1969, the squirrels were already extinct. I walked two transects through the beautiful ancient Scots pines in the main wood, where they used to live in the 1960s, before visiting Andrew Sutherland, the keeper at Glencalvie, who had worked there for seven years and never seen a red squirrel. He gave me a 4x4 tour of the estate and, again, I could see great potential for success.

The next morning my survey extended to the neighbouring estate of Croick. I met the head keeper, Alistair Sutherland, who had also never seen a red squirrel, but a visit to the main glen revealed some suitable woodlands. There was no doubt in my mind that it was another excellent location for a squirrel translocation, with the main woods in Strathcarron and Amat stretching out like fingers up the surrounding glens.

I prepared an application to translocate red squirrels and submitted it to Scottish Natural Heritage in September. In early December I again visited Alladale and Amat, to check pinecone availability, and found an above-average crop. We talked about nest boxes and feeders for a potential translocation in February, and the following day we received the green light from Paul Lister, with Alladale and the European Nature Trust funding the project. Once I received the licence from SNH, we started to plan.

Alladale manufactured a supply of nest boxes and nut feeders which I collected on 14 February. That day, I also went to Croick to meet the owners, James and Carol Hall, and with the keeper we went up to check the woods, deciding that the big conifer plantation on the north side of the glen was the best option. Between 18 and 24 February I put out nut feeders in a variety of woods near my home in Moray and in Strathspey, and started the process of encouraging squirrels to use the feeders. On the 25th and 26th I revisited them and attached the squirrel traps with the mechanism tied open.

27 February turned out to be a beautifully warm, sunny day. Following my pre-dawn journey to set the traps, Becky Priestley arrived at 9 am and we set off round the traps. Our first catch was seven squirrels and, after resetting some traps, we took them home and kept them in a quiet place in my garage. We went round for a second visit and caught four more, giving a day total of 11 squirrels, which we drove to the Strathspey Veterinary Centre in Grantown-on-Spey. Jane carried out our normal procedure of checking every squirrel for condition, weight, tail measurement and sex, as well as taking a blood sample. We had caught four males and seven females, all in good condition and with a good sex ratio of more females than males. Each one was placed, as usual, in a nest box containing hay, nuts and apples.

We arrived at Amat in the late afternoon and were met by the Alladale team, who helped us put up seven nest boxes with feeders in two locations at Amat woodlands, while the other four squirrels were taken for release at Alladale. Driving home, although it had been a long day, I was very excited that we had started another project to restore red squirrels to their old haunts.

The weather was cold and snowy in early March, so our next chance of trapping which fitted with the veterinarians' work commitments was on 13 March. Once again, Becky and I went round the traps, following my pre-dawn visit, and during the morning we caught 10 more squirrels, which we took to the vets in the early afternoon. Jane and her team were ready and we quickly worked through them: two of the females were pregnant. Becky drove them all north to Alladale, and she and Innes located five in the gardens at Amat House and the others at the far end of the Amat wood.

On 18 and 19 March, in very snowy conditions, several friends helped me put out traps in Strathspey. Fortunately the weather conditions improved so that on 20 March we were fortunate to catch 17 squirrels around Aviemore, Carrbridge and Grantown-on-Spey. This time Innes and Ryan, the under-keeper, came down in a 4x4 from Alladale and took the squirrels back to the release areas, including Croick. The success of these projects really depends on support like this, and from local people, who keep the feeders filled with peanuts during the initial stages of recovery. My next visit to the glens was on 23 April, when I could see that everything was going well: it was a pleasure to meet the locals who were now running the project. Once again, a wildlife translocation had created new friendships.

This population also thrived and numbers started to increase rapidly, as they had at Dundonnell. I remember a wonderful day when I called in at Amat House to see Johnny and Sara. As we sat at the table looking out over the river, we saw eight squirrels around their bird feeders, in nearby trees there were two more, and when I drove away there was one on the drive: 11, all told. Other interesting observations were reported: a couple living further down the river were looking out of their kitchen window when they saw a red squirrel dithering on the far bank of the fast-flowing river. Suddenly, in it jumped and started to swim. They dashed out of their house and ran down to the riverside, expecting to have to rescue it, but by the time they got there the squirrel was dashing up through the grass to the big pine trees by their house. We were learning that red squirrels, when they want to, can cross mountains and swim rivers.

Within a couple of years, they were seen 15 kilometres down the glen towards Culrain, and in the last few years have been recorded back in the woods between Rosehall and Bonar Bridge, on the far side of the River Oykell. We do not know how many there are, but there will be several hundreds or more living in an area of about 110 square kilometres, a mainly mountainous area with woods in the valleys, and they are still spreading. What a joy it is for local people who now have red squirrels visiting the gardens, and what a joy it is for us to know that we made it happen.

In 2012 I recognised that the red squirrel translocations were meeting our expectations and more, and decided that it was important to start moving on with the strategic vision of restoring red squirrels to all suitable forests in the north and west of the Highlands. I wrote an initial next-stage report, containing a map of the 25 best sites in Sutherland, Wester Ross and Lochaber, which I discussed with SNH, and submitted a fuller version in June 2014. I also foresaw a problem, though: a stepping-up of one of the requirements attached to the licences, the one that called for screening for potential diseases. This requirement had led us to start using a protocol, devised by Anna Meredith, by which the squirrels were checked at the veterinary centre in Grantown-on-Spey, and blood samples collected and sent to Edinburgh. The key disease to screen for was squirrel pox, but there were no cases of squirrel pox anywhere in the north of Scotland and – crucially – we were not translocating the animals into areas containing squirrels but were moving them to squirrel-free areas. The procedure involved anaesthetising the squirrels for a short time in order to carry out examinations and take blood before allowing them to recover inside the nest boxes. Jane was expert in this technique and no animals were lost during this process, but it was extra work, involved setting a specific date for veterinary examination, and the laboratory diagnosis came after release.

Jane Harley, who is a very experienced veterinarian, could not see that the procedure was necessary. She commented that I knew a healthy squirrel better than she did, in fact. The demand for screening became more problematic, though, when SNH mammal scientists said the animals should be blood tested for four more diseases, in addition to squirrel pox virus. It started to seem that we were being required to carry out scientific studies on squirrels rather than a conservation translocation. Fortunately, Ben Ross, head of licensing, recognised this and the matter was resolved by a letter of reference from Jane Harley, on 14 September 2015. It stated that she was confident that I was fully competent to carry out visual health checks and that she was always ready to help and give advice. This was an important step and allowed us to catch squirrels and release them that same day in their new location.

I know that SNH could see that the translocations we were carrying out were successful in creating restored populations of red squirrels in areas furthest away from any spread of grey squirrels. Meanwhile the work being carried out by the Scottish Wildlife Trust in preventing the spread of grey squirrels in Scotland and pushing their front edge further south was proving successful. At this time I applied for a licence for a five-year period which had a range of conditions which we discussed with SNH before any new location was chosen. In 2015 I started a new project at the famous Inverewe Gardens, a National Trust for Scotland property in Wester Ross, and then at Corrour in 2017. This was the start of a succession of new releases, mainly carried out by Becky Priestley. She was by this time working for Trees for Life, and I had recommended her for a similar licence because she had become a very experienced squirrel trapper and handler while working with me. After I'd given her advice on the first couple of locations she chose in Wester Ross, she was capable of carrying forward the recovery of red squirrels, and continues to do so. Interestingly, one of the first places she chose to release squirrels was in the pinewoods at Shieldaig, where, way back in 1999, I had been asked for advice by Hugh Tollemache.

I'm proud that we've really learned how to translocate red squirrels carefully, competently and successfully, with care and love for the squirrels themselves. It's interesting to reflect that we met so much opposition when we suggested the first projects, and that the translocation and reintroduction of red squirrels was not recommended, mainly for humane reasons: the animals, we were told, were very susceptible to stress-induced health problems, and while it might not be impossible to translocate them, it would be very difficult and with no guarantee of success. As it turns out, we have pioneered successful methods that can be used by others; in fact, I believe that we are now at a stage where the translocation of red squirrels should be standard wildlife management carried out by forest rangers, keepers and wildlife wardens.

I still find it baffling that the edge of the remaining range of red squirrels in the Highlands still seems to be eroding. Are there genetic impacts due to low numbers and closely related individuals, or are the numbers so low that they exhibit social problems, not breeding with closely related squirrels? Some day the 'new' squirrels will extend to these edge communities. I am also sure that translocation should be a tool in red squirrel conservation throughout the UK: remember that, one day, the species may become immune to grey squirrel pox and could then be restored to many more parts of England and Wales, especially where conifers dominate. I've thoroughly enjoyed working with red squirrels and making so many new friends throughout the Highlands, our paths meeting over an interest in this beautiful little mammal. This could, I'm sure, be replicated across the country. Despite the challenges of our early days, I just love the fact that people enjoy having them back – and as a country, we surely deserve to have them everywhere, and they deserve to thrive.

Beaver in the water, in Austria. (*Photo by imageBROKER/Alamy Stock Photo*)

5

# REWILDING

## *Bringing home the beaver*

This has been the most difficult chapter to write. The reintroduction of the beaver has been a never-ending saga, with very strong opinions on both sides. In fact it's been a bit like beaver country: the way ahead for beaver ecology looks clear, then suddenly there's a logjam, caused by farmers or politicians. The ideas run round the edge and create a new way through, but then there's another block, as one or other side of the argument triumphs and holds the whole thing up. From my side I could hardly believe the opposition's views – anyone would think that beavers will cause havoc throughout the countryside to all farms and forests. Anyone would think it's a herd of brontosaurus that was under discussion. In fact only a tiny percentage of properties countrywide would ever see a beaver.

My first experience of beavers in the wild occurred in June 1979, when I was on a RSPB sabbatical tour of Scandinavia to learn about avian predators at fish farms and the conservation of ospreys and goldeneye. For a few days that summer I stayed at the farm of Karl-Erik and Karolina Jonsson near Storfors in Värmland. It was a small, traditional family farm in beautiful countryside of native woods, mosses and water, with elk a regular evening visitor to their fields and beavers in their river. Cranes and goldeneye ducks bred close by, and several pairs of ospreys nested in stunted pines in forest bogs. On my first evening I walked along the edge of a flower-studded hay meadow to the river, which meandered sluggishly through deciduous woodland of birches, alders and aspen trees. Unfortunately it was a haven for mosquitoes. Stepping as quietly as possible along the riverbank, over beaver-felled birches, I saw my first beaver, but only briefly, for it sensed my presence and, with a loud splash, slapped its tail on the water as an alarm and disappeared. I sat down on a jumble of tree trunks, but my wait for further sightings was in vain, although I did get other glimpses later, and checked out my first beaver lodges and dams.

In July 1983 I returned to the farm on a family holiday, driving round Sweden and Norway. This time we saw beavers in the river and I got really good views at dawn of one on the riverbank. Again, I was fascinated by the beaver-felled trees lying in all directions, and the beaver-gnawed stumps. Karl-Erik told me that in winter he collected the dead trees, left by the beavers, with his tractor and trailer when the ground was frozen; they were nicely seasoned firewood and ready for his log store. He also occasionally hunted one, and he cooked us nicely braised beaver for our evening meal. It tasted to us like a cross between roe deer and brown hare. I saw other beavers and beaver dams on that visit, and I remember thinking again that this landscape was as Scotland should be.

In the early 1960s, and from 1970 to 1990, I worked full time as a field ornithologist with the RSPB in the Scottish Highlands, and it became ever clearer, as I looked at our land and read books on nature, ancient and new, that the absence of beavers and elk clearly diminished the ecology of our wetlands. The extermination in the past of the original Scottish fauna of wolf, bear, lynx, beaver, wild boar and elk had had a profound impact on nature. This came even more in focus to me in 1987 when the RSPB started looking at the purchase of Abernethy Forest, which represented a dramatic change for the Society and its staff, for this was a massive landscape acquisition rather than a normal bird reserve. It was an exciting project and involved many meetings about management in the early years, for it suffered from far too many deer, which prevented any natural regeneration of trees and plants. One of the first times I raised the return of the lost mammals was at a joint meeting of the RSPB and the Nature Conservancy on 7 July 1988. I guess it was another occasion when my friends thought, 'Where's he going now?' for while the need to reduce the numbers of red and roe deer was accepted by all, what I wanted in addition seemed like a pipe dream to them.

In 1990 I left the RSPB to start my own wildlife consultancy, so that – as well as working on bird conservation – I could also pursue these bolder ideas. I started to gather even more information on the missing species, like beaver, and read about their histories in the British Isles and their present-day status in mainland Europe. This was enhanced by discussions with a wider circle of colleagues, as in 1991 I was invited to be a board member of the recently created Scottish Natural Heritage, the successor in Scotland to the previous government agency, the Nature Conservancy Council. From 1992 to 1997 I was a member of the main board of SNH. On 6 May 1992 I was in a group examining a proposed mine in the hills of Perthshire. As we were hiking over the heather, the Chairman, Magnus Magnusson, and I got chatting about the missing mammals like beaver, lynx and wild boar. I quickly saw that he was thoroughly intrigued by the prospect of reintroducing them. That autumn I made a special journey to talk with the retired forester and naturalist, Don MacCaskill, at his home in Strathyre. Over high tea with him and his wife Bridget, I soon learned that he was keen on the return of the beaver to Scotland and had tried to encourage it, sadly without success.

During 1993 ideas about beavers continued to occupy my mind and I added a talk about restoring the lost mammals to my lecture itinerary. In May 1994 I arranged to meet Christoph Promberger, a mammal researcher at Munich University, and Professor Wolfgang Schröder, the team leader of the Munich Wildlife Society. We talked about the lost carnivores of Scotland but also ranged widely into ecosystem processes and reintroductions. My few days with them in the German Alps strengthened my determination to seek change. Public interest was also growing, and in September I did several interviews on beavers with journalists and even an out-of-the-blue late-night interview with Radio Canada. In late October Dr Vin Fleming, then head of biodiversity in SNH, asked me to give a talk on large mammal reintroductions to their team seminar at Forest Lodge, after which we had a lively and interesting discussion. Years later, Vin told me this was a pivotal point in recovery ideas. Beaver reintroduction was discussed at an SNH main board meeting on 1 November; it was a real mix of enthusiasm on the part of some members and antipathy from some science advisers. Restoring beavers, despite being ecologically easy, was going to be delayed by sociological and political lobbying.

I decided I needed to know more about the beaver in the field, so I travelled to the Netherlands in 1995. The Dutch biologist, Vilmar Dijkstra, studying them at the Biesbosch reserve in southern Netherlands, collected me by boat and took me out to their house and research site on an island, where we talked over lunch. I was told that six had been reintroduced in 1988, and that the beaver had previously been extinct since 1826. We went a short distance by boat and walked about a hundred metres through tall nettles to the first beaver lodge. There was lots of fresh evidence of mud and sticks, and during the summer the presence of young in this lodge had been proved by using a microphone on a spike, pushed into the inside of the lodge. There was a superabundance of willow growing along the water, so very little felling of trees had taken place – just a bit of bark stripping. There was so much plant material in this habitat that the beavers had very little impact. We continued through various channels, passing many pleasure boats, and came across a young female beaver in the water, just 4 metres away and not alarmed. I was shown three more beaver lodges, including one with only sparse covering, and I could see the two adults sitting in the chamber. They exited but stayed close, and were clearly used to people. I was told the male was 12 years old, with a young female, but they had not bred that summer. I was shown the first lodge where young were reared after the reintroduction. Vilmar showed me several places where beavers had left scent marks. I also learned that beavers only ate about 1 per cent of the annual production of plant material, and the flat, flooded landscape meant they did not build dams.

I followed Vilmar by road to Kekerdam, near the German border, where I met the beaver research expert Bart Nolet of Wagenen University, and fieldworker Freek Niewold. The beavers here were the second released population in the Netherlands and lived in a group of clay pits, with an incredible growth of willows and bramble and a rich range of flowering plants and shrubs; again, a superabundance of beaver food. It was, in fact, very like some of the big vegetated gravel pits of England. Bart's team were carrying out radio tracking studies of two beavers which were living in a home range of half a kilometre square. We listened to the signals of one very close to a busy public track to the river. The beavers were cutting down willow trees of up to five centimetres in diameter and this resulted in extensive coppicing. The 1994 released beavers had experienced some problems with leptospirosis, due to stagnant ponds and hot weather, and in the huge flood of the nearby River Maas in early 1995 the beavers had evacuated the site and swum to higher ground until the waters subsided. They planned to release more beavers in other parts of this catchment, so the animals could use the Maas as a highway. Before I left, Bart showed me the beaver way-marker pointing to the next beavers in Germany, and gave me a copy of his book *Return of the Beaver to the Netherlands*, inscribed 'Good luck with beavers in Scotland! Bart Nolet'. My knowledge of beavers was growing rapidly, thanks to the wonderful network of ecologists who were happy to share knowledge.

The next day, 24 September, I visited a German beaver research area on the River Spessart, near Hanover, with Sabine Hille, and her friend Mark Harthorn showed me his studies. Beavers were reintroduced to this area in 1987 and 1988, with 17 being released over two years. There were now about 100 beavers in rolling countryside with farms, woods and small rivers. We first visited a reserve at Sinntal with its special lazy-bed-like water meadows, which held much insect and botanical interest. The beavers had been building small dams and had created excellent habitats for birds and dragonflies. We then visited the small River Sinn, where beavers arrived in 1993 and built dams on the watercourse, just two metres wide, to create pools. They had created the most amazing wetland with alder trees cut up for dams. The main pool was now 70 metres long and 50 metres wide, with an island holding a lodge with beavers. It was a superb example of ecological changes due to beaver dams. Even small-scale changes, with small pools caused by their dams, had increased biodiversity: we saw two water shrews, birds such as snipe and grasshopper warbler were nesting in the marsh, willow tits made their homes in dead alder trees, and there had been big increases in plants like meadowsweet and kingcups. Dragonfly species had increased from 12 to 20. I was impressed.

As I stood there, I loved the light shining through the trees onto the beaver pool. Behind it, the tussocky meadow stretched across the small valley to a hillside of deciduous trees. In one pool there was a beautiful array of water lilies with brilliant yellow flowers poking out of the water, surrounded by the circular floating leaves, ideal for young frogs to sit on or hide beneath. In one area, the water came down through stunted Scots pines creating a series of pools with lots of cotton grass growing on the higher parts and bog bean pushing up along the edge of the river – it looked absolutely superb. At the far end of the pool the water was running over the vegetated dam, the speed of the flow making the river bed below gravelly, with the river sediment held above the dam. It's that mud that the beavers lift up with their front paws and plaster over the sticks which are used to build dams. Every night they check for leaks and dive down to pull up mud and roots to hinder the escape of water, for busyness is the key to maintaining dams and opening water channels. It's possible to use an excavator in a nature reserve to create pools, but after 10 years it will be necessary to go back with the machine again, whereas resident beavers work every night throughout the year, making certain that they have water deep enough for them to avoid predators and where they can build their lodges in the banks or on an island, as well as canals linking the different pools. They simply create intricate riparian networks, ecological engineering of the highest order, with multiple benefits to just about everything – fish, invertebrates, birds and mammals, as well as whole ecosystem gains.

We then drove into a valley leading into the forest, where the beavers had dammed a small river, which had allowed them access to a stand of non-native balsam poplar. The beavers used all of it, and after they departed the dam burst, leaving behind a beaver meadow. Finally, at dusk, Mark took us to a small lake with an island and, as the light faded, two adults, two sub-adults and three of this year's young came out onto the water. He showed me how to differentiate them, the adults just showing the top of their heads above water, while the youngest bobbed along buoyantly in full view.

On 27 September 1995, at the Zoology Department of the University of Halle, I met Professor Dietrich Heidecke, the expert on beavers of the River Elbe in Sachsen-Anhalt. He and his assistant Annett Schumacher showed me a range of beaver habitats, at one point stopping on the road where the river was diverted for coalmining activities. I could still see the old beaver burrows in an exposed bank. Dietrich told me that the beavers had been rescued and taken to Peine 20 years previously. We visited some ponds with beaver lodges and saw habitat management by them – I was told that three to four hectares of good habitat were sufficient for one pair. In Ragulf we watched beavers swimming under the town's road bridge, where they were feeding on willow trees hanging into water even as the traffic passed. One place on the Elbe looked rather like stretches of the Thames, but one arm of the river was so polluted that the beavers avoided it. The main river was clean, and just one kilometre was sufficient for one home range; there was also much evidence of non-native muskrats.

Further north, we arrived at a marvellous place where the beavers had dammed a stream and flooded several hectares of

Learning about beavers in Germany, September 1995 –
a well-constructed dam.

parkland, creating superb wildlife habitats. Our next visit was to a nature reserve where the beavers had created 10 long dams in recent years in riparian woodland. It was now the most excellent habitat for amphibians and fish, with the result that pairs of black stork, lesser spotted eagle and crane had arrived to breed. The first two species are very much followers of the beaver and its habitat creation. I could see that, by not having beavers in Scotland, many of our riparian habitats were impoverished, and we miss a lot. I was much impressed.

We then visited the Stekeby Biosphere reserve and its old hunting lodge, now used as a research station, where I met Peter Ibe, the beaver trapper, and his team. I saw the cage traps they used to catch and move beavers. The beavers of the Elbe were one of the few remaining European populations following intense persecution in previous centuries (other survivors were in Norway and on the River Rhône). The beaver was not killed because it was a pest; the very opposite was true. It was hunted to extinction because its fur, its meat and its oil were so valuable. The estimate then was 1,500 beavers in Sachsen-Anhalt, and Dietrich said that if we ever wanted beavers for Scotland, they would be delighted to help by sending families for reintroduction. After another great day with a renowned expert I really understood the animal and its needs, and how important it was to return the ecosystem engineer *par excellence* to our country as soon as possible.

The year following my experiences on the European mainland, I was able to reinforce my lectures on beaver reintroduction and was invited to speak at various events. Discussion of recovery projects at meetings of SNH now included beaver as a potential candidate for recovery. I also started to check various wetlands in the Highlands as potential beaver sites. For example, on 14 December 1995 I met my friend Andrew Matheson to look around his estate on the River Conon. We found many areas suitable for beavers with plentiful growth of willows, alders, poplars, bird cherry and sycamores close to the river, while nearby at Loch Ussie was suitable habitat for a family of beavers. There was no doubt that there were many rivers, lochs and marshes in Scotland already suitable, which was totally at odds with the views of scientists and others, who claimed that any reintroduction would be premature. Some conservationists said that beavers would seek out aspens, which they like, and cause the extinction of a scarce tree. The reasons against reintroduction were in fact social and political, not ecological, and most reactions to these reintroductions at large conferences, like the Scottish Ornithologists' Club, were overwhelmingly positive; with the message clearly being 'for-goodness-sake-get-on-with-it'.

Within SNH the beaver proposals were progressing very slowly; whereas Magnus Magnusson, John Lister-Kaye and I were positive, others were against the idea. Scientists were wary, while goodness knows what the Scottish Office thought. In 1997 my term as an SNH main board member came to an end, so I was no longer party to the thinking of government, although I was aware of the ever-increasing numbers of scientific reports on beavers, some by people who may never have seen a beaver, and growing public consultation on restoring beavers to Scotland. In 1998 SNH distributed a well-presented report – 'Reintroduction of the European Beaver to Scotland' – and invited comments from the public. The general population were in favour; in fact 86 per cent of written responses were in favour, and a further opinion poll of 2,141 members of the public gave 63 per cent in favour, 25 per cent with no view and only 12 per cent against. The latter were principally farming, fishing and forestry interests, and there was even opposition from some conservation interests because the beavers might 'annihilate' aspen trees.

Meanwhile, three beaver experts from Norway visited Scotland in 1998 and considered that a reintroduction would lead to viable but fragmented populations. A potential research area for a release had been identified in Knapdale Forest in Argyll and this led to local drop-in days. The beaver officer in SNH, Martin Gaywood, worked hard to bring many different strands of a potential reintroduction together, for it was proving contentious.

In autumn 1999 I was at one of the WWF large herbivore seminars in the Russian Caucasus, and friends from the Netherlands and Germany asked about beavers in Scotland: they just could not understand the delays. On my return home I wrote to Roger Crofts, Chief Executive of SNH, explaining my embarrassment at international meetings and asking if I should apply to SNH or to the Minister for a licence to reintroduce beavers to northern Scotland, if they were not going to get on with it. I explained that while SNH had been investigating the feasibility of restoring beavers to Scotland, five more European countries – Denmark, Hungary, Croatia, Belgium and Bulgaria – had reintroduced them. I explained that a local beaver group could carry out the work, find the money and had offers of animals from mainland Europe.

In November 1999, at the invitation of Olivier Rubbers, I attended a Beavers' European Day conference in Belgium. On 26 November I travelled by plane from Edinburgh to Brussels, meeting Peter Colleen, from Pitlochry Fisheries Laboratory, at the airport, and at Brussels airport I met Göran Hartmann, the famous beaver expert from Sweden, before the organisers drove us across the country to a residential outdoor centre called Ferme des Castors. We had a pleasant evening talking beavers, and our hosts Olivier Rubbers and Achille Verschoren explained that the old farm buildings had been turned into an outdoor centre for underprivileged children. They recognised the farm was named after beavers, which were extinct in Belgium, so they decided to do something about it. In 1997 they arranged for an expert to tell them about beavers, they spent weekends on feasibility studies in the Ardennes in the autumn and organised another weekend in October when about a hundred young people distributed 10,000 leaflets in the villages of the River Viroin, led by three drummers. The children sang about the beavers coming home, while wheeling a two-metre-high wooden beaver in a wheelbarrow. They put in an application in November but it was turned down the following summer. The following winter beavers started to be seen in the Ardennes – people must have moved some from Germany.

On the Saturday there was a series of good lectures by beaver experts from Sweden, Denmark, the Netherlands, France and Germany. It was a really interesting day in the village hall, opened by the local mayor. I met Gerhard Schwab for the first time: he was the wildlife biologist responsible for beaver management in Bavaria. On Sunday we had a field visit to the Ardennes, where we walked in steep-sided valleys of the River Viroin. The first place we came across was a superb area of beaver dams; the river was up to three metres wide and pretty straight, but the dams had created several small ponds, between 10 and 20 metres in length, with small canals, gravelly areas below the dams and backwaters of trapped leaves. The beavers had mainly cut birch up to 10 metres from the water. There was some oak, willow, alder and a couple of small spruce, as well as much meadowsweet. Since the beavers had changed the river, a pair of black storks had moved in and built a nest. The black stork is a rare, secretive bird which prefers small rivers and wetlands in forests, where it builds a big nest in the fork of an ancient tree. The beaver's fortunes have always been linked with the black stork. Both species once lived in the British Isles.

By now it was slow going, and in 2000, somehow, we guessed the beaver story was far from over and that there would need to be continual advocacy. I had suggested that some of us who were really interested in beavers should set up an informal group, the Scottish Beaver Network, to keep in touch and put out supportive information. The first meeting at my home included Kenny Taylor, Niall Benvie, Hugh Chalmers, David Jenkins, Hans Kruk and Duncan Halley, a Scottish wildlife scientist working in Norway. Duncan was a corresponding member of the network and a most valued contact for answering questions on beavers and for giving sound advice. Over the years, Duncan sent the most interesting information on beavers from throughout the world and was one of the great advocates of restoring beavers to his home country. On 5 March Kenny Taylor, Hugh Chalmers, John Lister-Kaye and I met with Paul Ramsay at his home at Bamff in Angus and, after discussions, looked at a wet area of 13.5 hectares with willows, bird cherry trees and lots of docks, meadowsweet and water cress in old ditches, and a rather sterile manmade pond. It looked a promising place to start a beaver trial.

A year later Derek Gow arrived to stay at my home and we had a great evening talk with him and my son-in-law Erik, a mammal expert from the Netherlands working in southeast Asia. I had first met Derek at an early species reintroduction and ecological restoration conference in October 1998 at Farnborough, and he has been a great friend ever since. Derek would become the greatest champion of the beaver. At our meeting next day, we also had with us Allan Bantick, who had set up a website for the Network, a very new thing in those days, which increased interest in our ambitions. One man even wrote from Lake Superior saying, 'Glad you guys are out there fighting on behalf of beavers. They are such wonderful creatures.'

We started to explore further the idea of setting up some demonstrations of beaver ecosystem activities within a secure enclosure, so that people could visit and see the importance of the species to nature, and that they were not the dreadful animal that some thought. Later we heard from a reliable source that the main board of SNH – by this time I was no longer a member – were meeting at their Battleby office near Perth to make a final decision on beaver reintroduction, and it was likely to be negative. By the year 2000, government had decided that meetings of public boards should be open to the public so, on 7 March, Paul, Hugh and I attended with notebooks poised. Martin Gaywood introduced a substantial document which recommended an option that SNH and Forest Enterprise should collaborate in a field trial. Roger Crofts emphasised the importance of the beaver for restoration of wetland habitat, especially in a time of climate change, but the debate was at times pretty awful: there was an 'anti' feeling, and too many 'not in my time' comments, while the chief scientist favoured more science. Fortunately the Chairman, John Markland, was in favour and the vote was just won in favour of beavers. On the way out, Ewan McIlwraith filmed an interview with me for BBC News, beside a small local river which was perfectly adequate for beavers. Later in the autumn, I wrote to the Minister, Sarah Boyack, Roger Crofts and the SNH Chairman expressing support and urging action; John Markland wrote back, saying, 'There is still a long way to go but we are continuing to move in the right direction!' How true.

In September 2000 Colin Galbraith, Head of SNH's Advisory Services, authored a press release stating that Knapdale Forest in Argyll was the favoured place for a possible trial reintroduction of beavers. It was hoped to release 12 beavers and the field trial would run for five years. In the press release, Colin said, 'We hope that over time beavers will help raise the profile of Knapdale as ospreys did for Speyside.' As an Argyll-born scientist it was nice that the beavers might be released in his native county.

In England, Derek Gow and I had discussed the impasse with John McAllister of the Kent Wildlife Trust. Derek was working at Wildwood in Kent, and we all decided to try to bring in some beavers for demonstration purposes to show their importance to water ecosystems. The ideal place to run a research project by the Kent Wildlife Trust and partners was at Ham Fen Nature Reserve, south of Sandwich, while a pair would also be available for Scotland to demonstrate the ecological benefits of beavers to riparian habitats. By the end of the year John had secured approval from English Nature and the Environment Agency to conduct a trial at Ham Fen. Derek had quarantine facilities at Wildwood, so in February 2001 Derek and I wrote to Erik Lund in the Department of Nature Management in Trondheim with an application for live trapping and export of beavers. I already knew some of these people through our white-tailed eagle projects, for that department in Trondheim licensed our collection and export of young eagles. A licence for 10 beavers was issued on 26 February and Derek organised the import of nine beavers to Kent, so the Ham Fen project could start. Then began a long wrangle with government about the project (in the meantime, two of the female beavers had given birth, each having two kits). The DEFRA view was that it was illegal to release beavers, while Kent Wildlife Trust said they were to be contained within a securely fenced site. It was a long and fractious exchange, but the wildlife trust was, in the end, able to start the project.

In Scotland, the Beaver Network considered the best place to have an experimental controlled beaver site was at Paul Ramsay's estate in Angus. In mid-October 2001 Allan Bantick and I drove there to see the new lochan and the beaver enclosure. We found an ideal place to build a lodge and Paul explained the plans for electric fencing. We were back on 4 November and it took us four hours to dig an underground beaver lodge, with the underwater entrance blocked with a willow sapling gate which the beavers gnaw through to get out after they'd been placed in the lodge. The pair of beavers arrived later and settled in very well, but some time later a tree, which was being felled, killed one of them – what a tragedy for an animal that should know how to avoid falling trees! A replacement was found and the beavers thrived. It proved to be a very important site for showing people the ecological benefits, and I remember Dick Balharry being very impressed. Others walked that path to those animals and their descendants, and it opened many people's eyes. I'm always impressed when I go back and look. On the last occasion a young man was starting a PhD on them, and Paul told me that he was the third doctorate student to research his beavers.

Meanwhile SNH drew all their information together and applied for a licence from the Scottish Government to import Norwegian beavers to carry out a trial reintroduction in a Forestry Commission forest at Knapdale. The trial would include an exhaustive research programme to test all manner of concerns about the impacts of beavers on the environment. Finally, after about three years, it was refused, with SNH asked to carry out further consultation and explain a range of issues of concern. I think this knocked the wind out of SNH's sails, and the project, like so many others, ended up in the long grass.

Dick Balharry and I were on an advisory group, which met regularly to give advice to the Highland Wildlife Park near Kingussie, part of the Royal Zoological Society of Scotland. On 1 May 2007 Dick and I were present at a meeting to discuss what should happen with beavers. The group included David Windmill and Iain Valentine of RZSS; Simon Milne, Chief Executive, Scottish Wildlife Trust; and Paul Watson of the RSPB. I knew that Iain was keen to see some movement, for we had talked together earlier in the year at Edinburgh Zoo. The meeting was very positive and we had a good discussion about the way ahead. It was felt that a partnership would give new impetus to the reintroduction of beavers, and after a good thrash around the subject, it was thought there should be two or three release sites, including one in the Cairngorms on the River Spey and the other at the SNH-preferred site at Knapdale. The ideas progressed well, and by October we held an open meeting in the Cairnbawn Hotel at Lochgilphead. Simon Milne of SWT had brought along his colleague Simon Jones, who would be the beaver officer, and his chairman Allan Bantick, with Iain Valentine from the zoo. Simon and Iain gave presentations and then there was a long question-and-answer session. A member of the SNH regional board and local landowner, Robin Malcolm, was very much against the project, but generally people were interested and supportive. This led, in early December, to a meeting of the beaver steering group in Perth, which was joined by Nick Purdy from Forestry Commission Scotland. We had a good, businesslike meeting and things were clearly moving forward.

On 30 June 2008 the steering group held a meeting at the SWT's headquarters in Edinburgh, with Martin Gaywood and Dave Batty from SNH. Martin was in charge of beaver issues within SNH, and Dave was the manager for the Knapdale area. The main discussion was about licensing, and it was a good,

The first release of beavers at Knapdale Forest on 28 May 2009, with a group including the Scottish Government Environment Minister.

constructive meeting which allowed the RZSS and SWT to get on with organising beavers from Norway and the subsequent programme of conservation management and research at Knapdale Forest.

Finally, a year later, beavers arrived and I drove down to Kilmartin in the evening and stayed overnight, which gave me a chance to catch up with Derek Gow and Gerhard Schwab. The next morning was the big day. There were certainly plenty of helpers and probably the biggest number of people ever to go to some of the lochs in the Knapdale Forest, for the Scottish Environment Minister was present to see the releases, along with the staff of SWT and RZSS, as well as people representing other groups and interests. This was the start of a research study which would go into just about every aspect of beavers in the wild at Knapdale.

I'm not keen on crowds, so a personal visit to the site in mid-March 2010 was so much more to my liking. At one point I walked along the forest track until I found where it had been flooded by a big beaver dam. I climbed up the hillside through scattered birch trees to skirt Loch Dubh and, once I'd got to a good viewpoint, I sat down in the heather. There was a beautiful vista before me across the sea to distant mountains, the Paps of Jura, but it was the noise below me that really drew my attention. The beavers had created an amazing home for what seemed to be hundreds of frogs; the chorus from them was awesome in the quietness of the late winter morning. My mind wandered briefly and I wondered whether they were singing a song of thanks to the beavers for providing such a wonderful new home. It had been hundreds of years since the last beavers provided homes for amphibians in Scotland.

Before leaving, I had a chance for a talk with my friend Roo Campbell, who was leading the research project. A good number of years before, when he had just left university and was at home in Inverness, he came and spent a few weeks with my Foundation, getting conservation experience. It was also good to meet Simon Jones and hear how he was getting on with leading the Knapdale beaver project. I was impressed that the project was underway but disappointed that the research had been skewed to looking at what problems beavers might cause and not enough on the ecological benefits of beavers – the increase in frogs, for example. I also realised that my worries, a decade earlier, that Knapdale would not answer the arguments about salmon fishing and farming were justified.

The Knapdale project continues, although there has been a need to add more animals, for it is not prime beaver habitat. The nearest enclosed beaver experiment to me is at Aigas Field Centre, north of Inverness, established by Sir John Lister-Kaye and his staff. There are excellent hides, overlooking the loch, for viewing the beavers at dawn and dusk, where thousands of visitors, including children, have learned about beavers and been shown their activities: felled trees, gnawed woods and their lodges. It has proved a very special wildlife experience for many. In Perthshire, beavers escaped from captivity and started breeding in the wild; the first escaped beaver featured in the Perthshire papers in January 1996, when it was killed by a motorist on the road near Killin and the police appealed for information. Others built up a population, which has been the subject of much argument and debate; on the National Species Reintroduction Forum we have had some heated exchanges with people in favour of removing them all from Scotland. But the beaver is a keystone species which will be essential for the ecosystem in a fraught future of climate breakdown. Farmers complain of damage to farming operations, but others say that it's the farmers' fault for growing crops within metres of rivers. I'd go further and say that every river requires a wide band of trees along its banks, to protect the river and prevent wash-off of farm chemicals and soil. With so much at stake environmentally – floods, droughts and the pollution of rivers and estuaries, which beavers can ameliorate – the opposition to beavers being a widespread accepted feature of our land is becoming ever harder to justify. In my view they should be restored to all water catchments.

* * *

In January 2005 Derek Gow and I visited Lake Vrywny, in Mid Wales, to meet Andy Warren of the local water company and we found many places suitable for beavers; Derek, in fact, thought there was room for four or five families. Later, in July, I collected Duncan Halley, from Norway, in Birmingham to drive to Newtown, in Powys, where we both gave lectures at the Welsh beaver conference, and where we met Derek and Andy again. It was a really good day with about 40 people from all walks of life. There were good discussions and some enthusiasm for trying to do something in Wales. In October Derek, Andy and I had two days with the experts in the Netherlands and Belgium checking out beavers and their ecological benefits, and talking with the people who lived near or worked with them. A week later I met Derek at the Cotswold Water Park where he was helping to create a breeding group of beavers in an enclosed gravel pit. There was tremendous press and media interest. We both lectured at a conference at the Eden Centre in Cornwall and then, in March 2008, I went on one of Derek's beaver field tours of Bavaria organised by Gerhard Schwab, the beaver professional. We were a mixed group of interests, and Gerhard and Derek showed us so many aspects of beaver ecology and management; we spoke with many people on our travels. I remember Gerhard's words of wisdom: 'If a farmer complains about beavers, be there the next day to investigate and talk. Do that, and about 90 per cent of complaints disappear. With the others, lend them a beaver trap. They may not use it but another 8 per cent disappear. Then you can deal with the difficult cases.' Those two experts are still running their fact-finding trips and have enthused hundreds of people about beavers.

Author holding live beaver in Bavaria in 2008, when on a study tour with Gerard Schwab and Derek Gow.

Beaver talks and lectures, meetings and reports really were the order of the day in that decade. Some really good examples of beaver research in controlled situations were started, and I remember walking in the rain with Derek around a reservoir in Devon, surrounded by farms, the topsoil running orange into the water. What a disaster. He then took me to the start of a beaver trial on a Devon Wildlife Trust site. A couple of years ago I was in Exeter at a conference when an Exeter University researcher gave a lecture about that trial. They had scientifically proven all the benefits that Derek and I had pointed out at the start. The dams slowed down the flow, evened out the floods, captured the silt and even the farm chemicals, maintained pools in times of drought and offered many other attributes. Derek has maintained enthusiasm and commitment for the beaver as an ecosystem engineer, and it's bizarre that the animal has not, yet, been restored to all water catchments. I'm sure it will.

Male polar wolf approaching the author's camp, Hold-with-Hope
in Northeast Greenland National Park, July 1988.

6

# RISK-TAKING

## *Large mammals and ecosystems*

When I was younger and lived in Strathspey, I really enjoyed the deep snow of colder winters, especially when it was perfect for cross-country skiing. One February in the 1980s, I skied from my home near Loch Garten through Abernethy Forest, crossing the frozen River Nethy, and then headed back via Forest Lodge. It was a gorgeously clear, sunny day with a sharp frost. The scenery of ancient Scots pine and juniper was outstandingly beautiful as I skied between the great trees. On my six-hour trail I saw a multitude of red deer tracks, lots of smaller roe deer slots and the trail of one fox. But, clearly, there was something missing from this wonderful forest – the tracks of a wolf pack following the deer, or a solitary line of pad marks of a lynx. The forest was, in some ways, ecologically dead.

As I skied, I passed a place in the birches called Lagg a'Mhadaidh, a Gaelic place name for 'the hollow of the wolf'. It was a reminder that these woods were once the home of exciting and influential large mammals. When I started to explore the Scottish Highlands as a professional field ornithologist, I used to come across place names recalling the mammals which had been exterminated, and stories about them. A favourite book for understanding these changes was James Ritchie's monumental work *The Influence of Man on the Animal Life of Scotland*. For me, it was a shocking account of the human onslaught on wild nature and the removal of the big mammal predators and herbivores over the past two millennia.

The brown bear used to be widely distributed in Scotland and was still present during the Roman occupation, with reports of Caledonian bears being taken to Rome to be tormented and killed in the Colosseum. The last were probably killed in the 9th or 10th century. The lynx was widespread in ancient times and was finally exterminated less than a thousand years ago, but its secretive nature meant that its disappearance went almost unremarked. Wolves managed to last much longer, despite intense attempts to kill them, with the 'last' wolf reported to have been killed near Tomatin in 1743. The elk, the Irish elk, aurochs (wild cattle), wild boar and beaver were all vanquished from their ancestral haunts as human impacts on Scotland increased. This assemblage of species was, in fact, essential to the functioning of the ecosystem, and once they were gone, Scottish nature changed.

I first visited the European mainland in the early 1970s and remember seeing the signs of wild boars digging in forest clearings in the Netherlands and France. In the summer of 1979 I watched beavers on a sluggish river in Swedish Värmland and massive elk in Jämtland, where a local zoologist told me of brown bears in the nearby wooded mountains. I was starting to feel guilty about what we were missing – about what we had exterminated – in Scotland. That recognition deepened on my first visit to the primeval Bialowieza Forest of Poland in 1985, where I recognised that we had to think very differently, and boldly, about nature conservation in the

Scottish Highlands. But how slow that journey has been, and what a lot of social and political obstructions have thwarted progress.

In 1987, while I was the RSPB's regional officer in northern Scotland, the owners of the major part of Abernethy Forest in Strathspey offered to sell this fantastic natural area to the Society. We were excited but, at headquarters, there was caution, due to the cost and the scale of what was on offer. We came to a crunch point, so Dr Ian Prestt, the then Director, came north to help himself make up his mind. I remember taking him up to a viewpoint near Rynettin croft, which overlooked this superb landscape of Caledonian forest surrounded by mountains. As we stood there looking out over the ancient pines, Ian asked me how big the reserve would be. 'Everything we can see and more,' I replied. He was so excited by this incredible, thrilling prospect that it encouraged him to confirm the decision to complete the purchase. The RSPB gained much more than a then-normal nature reserve; it took a step into a landscape future. The first major task was to reduce the over-population of red and roe deer to allow ecological restoration; in fact I was thrown in at the deep end, as it was my task to explain to the RSPB Annual Members' Conference in York why we needed to kill a lot of deer. But I could also see that, with such a large reserve, the RSPB needed to think about using lost mammals to restore a functioning ecosystem. In our discussions in 1988 I raised these ideas – as a starter we would need beavers, we needed wild boar, we needed lynx, we needed wild cattle. And then we needed to return the wolf, the elk and the bear. Wouldn't that be fantastic? Sadly, at that time, it was all a dream too far. It still is, for many, even for many conservationists, and all those keystone mammals are still absent from Abernethy Forest. I feel I failed.

The idea of conserving, enhancing and expanding our best nature areas and reintroducing the extinct mammals was slowly starting to attract attention in the 1980s, but it was going to be a slow progress. There were some voices in favour. In 1986 David Stephen, the Scottish naturalist and writer, published *Alba, the last wolf*. He kept a pair of wolves in an enclosure at Palacerigg Countryside Park, near Cumbernauld, and was an advocate for the species' reintroduction, while the forest ranger Don MacCaskill was a great advocate for the return of the beaver. Others also hankered to see things change, but there was strong opposition from farming, forestry and land-owning interests, and even from conservationists. It's still the case today.

I became ever more interested in species reintroduction and ecological restoration, and was able to follow this inclination more strongly when I left the RSPB in 1990. I became an independent ecologist and set up the Highland Foundation for Wildlife. By one of those strange chances, an investigation into two German falconers stealing peregrine falcon eggs near Inverness, while I was still with the RSPB, finally came to court in the German city of Mannheim on 10 May 1994. A police officer from the Northern Constabulary and I were called to give evidence at what turned out to be a successful prosecution. Earlier in the spring I had read an interesting article about large carnivores written by a researcher at Munich University. So before going to Germany I faxed a letter to him asking to meet and discuss the missing mammals of Scotland. Christoph Promberger replied very positively. 'Come and see us at our mountain field station,' he wrote, so after the trial was over, I hired a tiny car and drove 400 kilometres to Garmisch-Partenkirchen. I was very fortunate to meet Christoph for the first time and the team leader, Professor Wolfgang Schröder of the Munich Wildlife Society. They found me a place to sleep and, talking with them over a barbecue and through the next day, I learned a huge amount about wolf, brown bear and lynx. We became firm friends and I now had knowledgeable and enthusiastic experts on whom to test my ideas and who could encourage me in my aims.

In the early 1990s, after this visit, and from reading widely and walking through Highland forests near my home, sometimes in the company of my cattle, I became even more convinced that large mammals were an important part of the ecological restoration of the best nature areas of our country. In fact, for the future of nature and humans, I saw that there had to be a much better balance between modern farmed lands, the uplands overgrazed by sheep and the natural ecosystem areas. My initial view at that time was that at least one third of our land, and not just the impoverished uplands, should be exclusively for nature and its many benefits to ecosystem health and the future of the planet. Now I think that half our land and seas need to be dedicated to nature, for the damage we have done has been profound. I soon realised that it was not just a matter of bringing back charismatic predators like wolf and brown bear, which was often the first thing to spring to people's minds: an important requirement was the ecological impacts of the large herbivores, like the beaver – as we saw in the last chapter – and also of the bigger ones, such as elk, wild boar and aurochs, the latter species extinct but partially replicated by native cattle. Their importance, especially in wooded and marshy landscapes, sprang from their grazing and browsing, creating ecotones in the vegetation, glades and open areas, while their heavy, trampling feet provided tracks and pathways for smaller creatures and their dunging dispersed seeds throughout their range.

Another step forward in my thinking came when I went to stay with Christoph Promberger in Transylvania, where he and his team were researching wolf, bear and lynx. I remember vividly the occasion when he drove me to meet Mosu, the local gamekeeper; in the late afternoon I walked with them to a hunter's tower in the forest. These wooden towers had been built by the Romanian State Forest Service for bear hunting, although it hardly deserved the name of hunting, as bears were encouraged to come to feed on maize and fresh meat. The deposed president Nicolae Ceaușescu had been a fanatical hunter of bears in Communist times, so bear numbers were artificially high, thanks to supplementary feeding and protection. Great beech trees cast dark shadows across the snowy clearing in the forest. At just before six, two young brown bears walked in from the forest and squabbled over a feed of maize. To me, they looked huge, but my Romanian companion told me that they were only half-grown. He thought they were siblings and, to judge by the way they behaved, he thought a big adult was somewhere nearby, waiting in the wings.

Mosu's beat was about 13,000 hectares of state forest, encompassing three valleys with steep wooded sides as well as a few small farms by the rivers. Ten kilometres away, in a broad fertile valley, were towns and villages, while Brasov, a city of 250,000 people, was only 30 kilometres distant. As we sat and watched the night creep in, we talked of nature. Christoph translated for Mosu and for me. Mosu knew exactly what was in his forest: there were 105 red deer, 120 roe deer, 160–80 wild boar and 43 brown bears, as well as six lynx and two packs of wolves, totalling five to seven animals in each pack. We had spent several days tracking some of them in the deep snow, but had seen only two red deer and one bear in daytime.

Mosu was curious about Scotland and found it difficult to believe me when I said a similar-sized area in the Scottish Highlands might hold 2,000 red deer and 400 roe deer. I said that we had killed all our bears, wolves, wild boar and lynx centuries ago and that all were extinct. I tried to explain about overgrazing by red deer and the lack of tree regeneration, but I'm not sure he understood, because his woods showed excellent regeneration. With a big smile, he offered to come and reduce the deer for us. If he ever had, he would have been amazed by the difference between his and our red deer, ours being probably half the size of his because of long-term degradation of their habitat. Finally, it was dark and we climbed down from our lair and drove off into the night.

Some years later I was invited by Magnus Sylvén of WWF International to join the recently created Large Carnivore Initiative for Europe. We had the most stimulating meetings to discuss future conservation and research requirements for brown bear, wolf and lynx in Europe. I met some great people and learned much, for I was the only one present from a country with no large carnivores. Subsequently, Magnus suggested the establishment of a Large Herbivore Initiative for Europe, and that again involved stimulating meetings, discussions and field trips, with occasional joint gatherings. At home, I advocated the reintroduction of a range of lost species, and although this was well received by many and warranted excellent coverage in national newspapers at the time, I realised that it was going to be a steep uphill battle with many opponents, for environmental as well as social and political reasons. At that time I was on the main board of Scottish Natural Heritage and regularly championed these causes, but although I helped stimulate the debate, I was unsuccessful in helping get the mammals back home. Our horizons were too small and, sadly, they still are, although recent activities and thinking on rewilding are encouraging signs of change. Never believe anything will happen until it does, though, however encouraging the signs – for whispered conversations in dark corridors can always block anything, even at the very last moment.

## Wolf

Whenever 'restoring the wild' is talked about in Britain, it's always the supposedly dangerous wolf that is first to be mentioned. I suppose that makes sense, because it was the last big mammal predator to survive here. It's incredible to think that, despite the trapping and the poisoning and the general hatred directed towards it, the wolf was the one which lasted the longest. Yet here we are, a nation famous for its love of dogs, which have been around humans for thousands of years, ever since we domesticated them from – yes, that's right – the wolf, so our history with them is long and complicated. Wolves were, and still are, widespread, all the way across Europe, Asia and North America and into North Africa. In Britain they were widespread too, until they were slowly driven north, the last ones surviving in the mountains of Scotland. 'Last wolf' claims are legend: there used to be a stone marker beside the A9 trunk road at Loth, north of Brora in Sutherland, to commemorate one of them. I remember once going with a keeper on his Argocat over the hills to view a jumble of massive rocks – that region's last wolf lair, at the back of the mountain. There's another one near where I live in Moray, where wolf cubs were said to have been killed in their den, while just a little further away is the Findhorn River, where in 1743 a hunter named MacQueen dispatched with his dirk – a Scottish knife – what was said to be the very last wolf in Scotland. Far more likely is that the last wolf in Scotland crawled away on its own and died in some lonely hole in the rocks. The last of Scotland's big predators was lost, following the lynx and the brown bear into oblivion.

References to wolves occur throughout the country in place names, historical traditions and writings and legends about wolf hunts, as well as in carvings like that of a wolf on a Pictish standing stone near Ardross in Ross-shire. An old farmer friend in Strathspey, now dead, showed me a wolf cot, a stone enclosure which his ancestors used as shelter for their few cows, sheep and goats, protecting them from night-time wolf attacks. There were several others in that district, too. Much later, I saw very similar stone-walled pens topped with a barricade of pointed sticks built by present-day Romanian shepherds in the Carpathian Mountains to protect their stock. Ancient wolf traps or pits have been identified in various parts of Scotland: the onslaught was intense.

I saw my first wolf in 1988 when I was on a goose research expedition with the Wildfowl Trust to the Northeast Greenland National Park. The Otter aircraft we were on was forced to land on a beach a mile away from our base, so we carried our gear, back and forth, to the old Norwegian hut at Myggbukta. In one patch of beach sand I clearly saw the massive dog-like prints of a wolf. Before going, I had read up on wolves, Arctic fox and musk ox and hoped to see them all.

We got on with catching and ringing pink-footed geese in order to study their journeys to Britain. Our camp was in the land of the midnight sun so, late on the fourth evening, I walked up the 180-metre hill overlooking our campsite in its broad glacial valley, for I was determined to see a wolf in the midnight hours. It was a beautiful night: no wind, no clouds, and the low sun turning the whole land golden, with behind me the calm Arctic sea scattered with small icebergs. I kept on scanning with my binoculars for half an hour. Close to me were snow buntings and a pair of ringed plovers, and I could hear the geese honking in the lakes below. Suddenly I saw what looked like a white rock, nearly a mile away on a hill slope north of me. It never moved, so I carried on scanning the whole valley. Twenty minutes later, there was what I might have taken for another white rock, until both stood up and started to howl, the midnight sun backlighting their white coats. I remember vividly how the hair on the back of my neck stood up. I was awestruck by the magic and mystery of the moment in the presence of howling wolves. Two quarter-sized greyish pups then scampered to their parents, after which I saw the male stalking off north to hunt and the female and pups walking back to their hidden den. It was one of the most special moments in my life.

Some days later two of us were doing a long transect for breeding wildfowl and waders, walking half a mile apart through the valley, when, at one point, I descended into a small stony hollow. As I went over the edge the female wolf ran out with her pups. I ran after them, hurriedly fixing a telephoto lens to my camera. By the time I got to the top of the other bank, she was leading her pups quickly up the slope, so I stopped, sat on a rock, cupped my hands and started to howl. She stopped immediately, all four of us howling, and then she slowly came down to investigate, walking a slow cautious circle around me at about 150 metres, looking at me continuously, then leading her pups away into the hills. Later in our expedition, we were sitting and talking over a very late breakfast after catching geese when we saw that the male polar wolf was silently walking towards us over the flower-dotted turf, the hill behind smudged with snow patches. I had read up on polar wolves: they were curious, unafraid – nowadays – of humans and not dangerous. All we had to do was keep still, but there was a scramble for cameras. His tail was between his legs, telling us that he came in peace.

He was taller and larger than an Alsatian dog, all white with black nose and mask, grey eyes and large feet. He walked around the campsite, passing within three or four metres, to the click of cameras. We felt no fear, just awe at this magnificent, much maligned animal, and pleased that we lived in an era when we did not reach for a gun. As silently as he arrived he walked away, leaving an indelible image in our minds.

Every now and then someone suggests the reintroduction of wolves to Scotland, and the debate is reignited, nearly always led by people who have never seen a wolf. I've always believed that before giving an opinion on a sensitive issue like a reintroduction, you should already know the creature you're talking about, and where it still lives in the wild. There has been some excellent research carried out on the potential numbers and impacts of wolves in Scotland, especially in relation to red deer. Those that have studied wolf reintroduction know that the necessary habitat and wild prey are present, but the whole debate revolves round social and political issues.

In December 1994 Professor Wolfgang Schröder and Christoph Promberger came to stay with me in Strathspey. As I drove them from Aberdeen airport over the mountains by the Lecht, they were shocked at the state of a land devoid of its original tree cover. We had an excellent three days in the field, first visiting Abernethy Forest, followed by a discussion with RSPB staff and wardens. The next day we visited Glen Affric and the other north Inverness glens with John Lister-Kaye, and on the last day we toured the Insh Marshes west to Creag Meageadh, Glengarry Forest, Glen Quoich and back along Loch Ness and the Monadhliath mountains. We talked all day and made plans over dinner at home; on the last evening we were joined by Dick Balharry and his son David, Peter Reynolds and Alan Hampson, all mammal experts with SNH, for a wide-ranging and thought-provoking evening. The parting message from my visitors, as I drove them back to the airport, was that Scotland had excellent areas, large enough for lynx, wolf and brown bear. They could see no ecological barriers to successful reintroductions; it was solely a matter of social and political will. Nothing since has changed.

Campaigns for large mammal reintroductions are a succession of ups and downs in Scotland, occasional hope always followed by despair, a rise and fall of unanswered questions. In March 1995 a meeting in Elgin was called by the founder of the Highland Wolf Fund, who lived in Oxford and wanted to reintroduce wolves in the Highlands. It was a well-attended and boisterous evening attended by a hundred people with a wide variety of views; I sat at the back, quietly listening with Dick Balharry. At times it went badly, especially when some Strathspey farmers suggested setting up an Oxford wolf reintroduction to release wolves into that part of England. The speaker did not get the sarcasm and there was no chance of his idea getting off the ground.

There was a very different evening in November 1997, when Christoph Promberger was invited to speak at the annual gathering in Inverness of crofting assessors from all over the Highlands and Islands. This might sound like an invitation to the lion's den. Christoph was by this time well into his Carpathian Large Carnivore research and was well known to the villagers and farmers in the mountains and forests of Romania. He was reared in a forest and farming community in Bavaria, and as a down-to-earth expert biologist he gave an illustrated lecture about his studies. His audience was amazed to hear that there were eleven million sheep living in the Romanian mountains. At question time he was quick to point out that he was a visitor and that it was not up to him to say 'yes' or 'no' about reintroductions: he just hoped to expand people's knowledge. Most of the audience kept sheep, but the questioning was really exploratory and sensible, which is what happens if a speaker really knows his subject.

I went to Romania several times to visit Christoph and his wife Barbara, who had joined him in his carnivore research in 1996, and I learned a huge amount. I failed to see a wolf there, but I did in northern Canada and, years later, on a winter visit to Yellowstone National Park, where Doug Smith, the wolf scientist, explained their reintroduction project and showed us the famous wolf packs of the Lamar valley. That visit was with Paul Lister, the owner of the Alladale estate in Sutherland. He had tracked me down in the late 1990s to talk about large carnivores, as his wish was to see them reintroduced, although his plan was for them to live in a massive enclosure. He was a wealthy man who had hunted red deer in Scotland before coming to conservation. I introduced him to Christoph in Romania before he purchased the Alladale estate, rebuilt the infrastructure, turned it into a major wildlife reserve and created new woodlands of nearly a million native trees. I liked and supported him. He was a landowner in a different mould from those whose staff killed protected raptors, but his plans were difficult to put into action, there were just so many obstacles in his path. I remember I spoke for him at a very well-attended meeting in Bonar Bridge, the village hall bulging with people keen to hear him outline his plans, which had attracted great media attention. It wasn't long before the older sheep farmers piled in with angry opposition, which I suppose was to have been expected, but what was encouraging was that, after questions, so many young locals came up quietly and said they would be in interested in working at the reserve. It would beat driving to and from Inverness every day for work, they said. Sadly, Paul's vision is still a dream.

In Europe the wolf has recovered its range and numbers, even living in well-populated regions. In recent decades it has returned to its original haunts in Germany, the Alps and France, and has even gone into Denmark, the Netherlands and Belgium. In late autumn of 2018 I was in the Netherlands for a few days to learn about their white-tailed eagles, and one night stayed with a friend living in the Veluwe Forest near Arnhem. He told us that a female wolf had recently walked from Germany and settled down not far from his home, for there were lots of deer and wild boar there, and now I gather that a wandering male has found her, and with luck will start a new Dutch population.

Do I think we will have wolves back in Scotland sometime soon, or at any time? I'm not sure. What I do know is that, unlike in other countries in Europe, they cannot simply walk here across our borders. If we want them, we will have to intervene. In the past, public opposition was great because of sheep farming and a fear of wolves engrained in a population that had not known them for 250 years. They could live here, though, and there is an urgent and massive requirement to restore the ecology of our degraded lands after centuries of overgrazing by sheep and deer. There have been dramatic changes in the way we live – we eat less and less meat – as well as societal changes in the way we farm, alongside an ever-increasing recognition that we must restore ecosystems to help combat climate breakdown. When we're thinking about the future of the earth and our grandchildren's grandchildren's future, the return of the wolf makes sense. Always remember, though, how long it took to reintroduce wolves into Yellowstone National Park, a wilderness with no sheep. The first proposals were mooted in the early 1930s: the eventual release came about in the 1990s. It takes time and sometimes more than a generation of biologists, but a failure to fight for it now will mean the wolf will be absent for two or three decades more.

## Lynx

Of all the species in this chapter, the lynx is the one which could be introduced most easily. It would contribute markedly to ecosystem functions by hunting and disturbing roe deer, as well as acting as a potential apex predator on middle-guild mammal predators such as fox, badger and marten, which are now very common. It would also restore an iconic and charismatic large cat, the size of a small Labrador dog, to its former Scottish range, which could bring real ecotourism benefits. The proposal has been talked about since the early 1990s – over the years, the feasibility research on the species has been of the highest standard – and it's a disgrace to UK wildlife conservation that the species is still absent.

I feel particularly disappointed because when we started to talk about lynx I really thought that reintroduction would not be far into the future. In 1988, when the RSPB purchased Abernethy Forest in the Scottish Highlands, this was a key species to interact with deer, especially roe deer, once the large numbers were reduced with the rifle. It's their stealthy presence, not just their hunting, which keeps the deer moving and allows the ecosystem to heal. The early years of Scottish Natural Heritage in the early 1990s was a time of promise and hope for nature, an era when we thought we could do great things for Scotland's wildlife and environment. Lynx featured in that optimistic thinking, and my interest in reintroducing lynx took hold when I visited Wolfgang Schröder and Christoph Promberger in Bavaria in May 1994, to talk about lynx, wolf and bear. That winter they came to stay with us in Strathspey to give their expert assessments on the potential for reintroducing large carnivores in the Highlands; the lynx was a no-brainer. Its reintroduction would be simple, if it was allowed. Early the next year I was very fortunate to spend nine days with Christoph in the Carpathian Mountains in Romania as he and his colleagues conducted field research on the three carnivores. I didn't see a lynx but I saw their superb pad marks in the snow and, one evening on a high ridge, was told that the call we could hear further down the slope was that of a lynx. I was captivated by the magic of this elusive animal.

In April 2000 I had a letter from David Hetherington, a recent graduate of Aberdeen University, asking if he could visit and talk about large carnivores. He was working at the Caledonian Partnership in Munlochy, where a mutual friend, Tim Clifford, had suggested he contact me. David had been at my lectures on reintroducing large mammals and had talked with me about lynx three years previously at Beauly. I suggested we meet in Inverness and I agreed he could come and do voluntary work on lynx with us. This finally led to David starting an MPhil degree and then a PhD at Aberdeen University, with my Foundation giving significant funds to carry out the research for the first few years. A little later that year, I was at a meeting of the Large Carnivore Group and was talking with the Norwegian scientist, Reidar Andersson. He told me about a recent experience when he took the Norwegian Environment Minister to a lynx den with young, as part of their research project on lynx, roe deer and red fox. I was envious and asked if there was any chance that I could visit the following summer. He said he would get in touch.

In late May 2001 I got the call from Reidar, telling me to get the plane to Oslo early on 7 July. I got off the flight from

Aberdeen to find Reidar and his son waiting at Oslo airport, and they drove me out into the countryside, not far from the city. Up a forest track we met his colleagues and sat drinking coffee for nearly two hours while, further into the woods, his colleague Lars worked out the exact location of the den by using the radio tracker on the mother. He phoned Reidar to say he was very close, so we walked up the track through a wet valley bottom, quite narrow, with rocky cliffs and screes on either side and much tree cover, mainly Scots pine and Norway spruce. We met Lars pointing to a big scree-covered slope of large rocks with lots and lots of holes. It took nearly an hour to locate the den, the female occasionally growling from the trees above. Then, suddenly, one of the researchers found the lair underneath a big rock and caught two five-and-a-half-week-old kittens. I noted every detail in my diary: 'Fantastic little guys, splendid speckled fur, velvety to the touch, typical ear tufts and big feet.' After biometrics, the veterinarians fitted a radio transmitter to one – both were males – and tattooed both in the ear, marking one 22 and the other 23. I had a fantastic opportunity to photograph them, and before they were placed back in the den I was thrilled to be able to hold one. In fact, I wished I could have popped them in the pockets of my anorak and taken them back to Scotland. They were placed back in their den, with the female nearby, and we all left.

That night I stayed at their field station with the team, enjoying a very nice barbecue and talking until late, with a quail calling from a cereal field. The next morning Reidar and his team were checking roe deer to work out which had fawns, and we saw a lot of elk tracks. That evening, he went searching for a male lynx, which he'd called Haakon, by driving on forest tracks and small public roads until he got the distinctive peeping from his radio transmitter. At times, Haakon was very close to us,

Author privileged to hold a lynx kitten before returning it to its den near Oslo, Norway, in July 2001.

even if we couldn't see him; as compensation, we had a very nice view of a female elk and later, at midnight, a red fox. The lynx spooked a roe buck very close to us, which dashed with a really long series of agitated barks, and managed to escape. Haakon moved quite fast to the west and crossed the road behind us, just as we had a spotlight on the fox in front of the vehicle. It was the most fantastic evening tracking lynx in a landscape of small farms and woods, not out in the wilderness but just like where I lived in Scotland. The suburbs of Oslo were not far away, and the team said that they could sometimes pick up the lynx transmitter signals in the hills from the centre of the city. I went home absolutely convinced that the lynx could return to the Highlands and live alongside us with ease.

In July 2002 the Swiss experts on lynx, Urs and Christine Breitenmoser, came to stay with us on the croft in Strathspey; this was another opportunity for them to look at the Highland landscape and talk about lynx. The following February, David Hetherington and I met Urs and his colleagues in Bern; we were enjoying a dinner in a restaurant in town when the phone rang to say that a lynx had been caught in a box trap in the French Jura. We set off through the night on a fast drive to Biel, through road tunnels to a village in the mountains and so to a hamlet, before setting off up forest tracks with Urs, and his colleague, Fridolin, pulling a sledge. We met a game warden at the trap and quickly had a look at the young lynx, growling away. It was another beautiful creature with great big paws, impressive ear tufts and beautifully patterned fur. They darted it and carried it down to the game warden's house where, on examination, it proved to be a young female. She was taken down to their headquarters for translocation to the Alps with her mother, which had been caught a few days earlier. The team's aim was to mix the genes between the two populations. I got to bed at 3 am after a fantastic day, and we enjoyed two more days, learning about lynx with the experts.

David completed his PhD in 2005 and it included everything one needed to know about the ecology of lynx for a reintroduction. We promoted the proposed reintroduction on the website of the Highland Foundation for Wildlife and I sent a letter to SNH requesting information on licence procedures for lynx, wolf and brown bear, but it was never answered. Lynx reintroductions were not an untested idea, for they had been successful in Switzerland, Poland and Germany. My visits to Portugal and

Spain concerning osprey reintroductions had allowed me to learn about the highly endangered Iberian lynx, a smaller cousin feeding principally on rabbits. The wildlife guards in the Coto Doñana National Park tried their hardest to show me one, although it turned out to be the typical 'You should have been here yesterday' kind of visit, literally so on one particular day, when we stopped by a trail camera, looked at the footage on the card and saw a male lynx casually walking past at that exact time, 24 hours previously. Later that evening I was permitted to visit the El Acebuche lynx breeding centre and watched the captive breeding lynx via CCTV. The Spanish and Portuguese have mastered the art of breeding Iberian lynx in captivity and

Overlooking the successful breeding station for Iberian Lynx in the Algarve, Portugal, on 12 September 2014.

were able to supply many for release. On 12 September 2014, while on an osprey project, my friend Luis Palma organised a rare and privileged visit to view the Portuguese breeding centre near Silves in the Algarve hills. The director was a fund of knowledge and I learned even more about this beautiful animal. We sat for a time on a hillside overlooking the 16 big cages, holding Iberian lynx of all ages. The recovery in Iberia is progressing well, but looking at those lynx I wondered, in a time of climate breakdown, whether the UK could provide a home for a safety population of Iberian lynx in the wild. It's worth considering.

From then on, more research was done, and in 2015–17 we held a few meetings with colleagues in the Scottish nature NGOs, but they led nowhere. In 2018 David published a beautiful book called *The Lynx and Us*. There was no excuse now: all the supporting material for lynx reintroduction to Scotland was in the public domain. On 14 June 2018 my Foundation submitted a reintroduction application to SNH, with a covering note saying that, following the successful start of our sea eagle reintroduction to the Isle of Wight, we were dusting down the proposal and having another try. One day it will happen, and I hope it will be sooner rather than later. In 2018 the Westminster government published a 25-year plan for Nature in England and Wales, and included lynx reintroduction. Like the wolves of Yellowstone, there's been enough research. At some stage politicians must be bold enough to show support so that we can just get on with it. Surely it's time for the Scottish government to license the return of the missing lynx?

## Brown bear

I most often think about brown bears in Scotland when I'm in the headwaters of big rivers, like the Findhorn and Dulnan, at salmon spawning season in late autumn. It's an endless cycle of salmon eggs hatching in the uppermost tributaries and travelling out into the ocean as small fish, then returning with the ocean's nutrients as big fish to spawn where they were born. Several thousand years ago and more, these very places where I now stand would have witnessed the annual harvesting of salmon by Scottish brown bears, wolves, eagles, foxes and other fish eaters. This is a spectacle still occurring in the great rivers of North America and Russia, where local people and ecologists recognise the benefits to the whole ecosystem of the salmon's return from the ocean, their bodies returning nutrients to the land. In our country, sadly, that link has been broken for millennia and, due also to deforestation, our river catchments are impoverished.

People find it hard to believe that the brown bear was once native throughout our country, associating it more with far-off places. Bear bones have been found and carbon dated, with James Ritchie considering that they were finally exterminated in Scotland a thousand years ago. The Romans regarded the Caledonian brown bear as fearsome and special, and enjoyed them as fighting animals in the Colosseum, while in Scotland they were clearly revered by local tribes, for their images appear on Pictish stones of the 7th and 9th centuries. While we might find the idea of fighting bears repugnant, we are clearly still captivated by the animal close up: people in Britain may remember Hercules, the famous brown bear born in captivity at the Highland Wildlife Park and raised by the Scottish wrestler Andy

Robin and his wife Maggie at their home near Dunblane. At one year old, he weighed 190 kg and appeared with Robin in a wrestling show from the late 1970s to the 1980s. On 20 August 1980, while filming a TV commercial in Benbecula, he escaped and disappeared for 24 days. A crofter saw him swimming between small islands in the Hebrides and he was eventually recaptured using helicopter and tranquilliser darts. It's amazing that he survived, unseen, for nearly a month in the wild, so maybe he proves that there is still room for bears in Scotland. Back in captivity, though, stardom followed, even including two years in California, until he died of old age on 4 February 2001. In 2013, a life-size statue of Hercules was erected on North Uist.

That Hercules the entertainer is the only bear known to many in Scotland maybe tells a story of its own. I saw my first brown bears in the wild in Romania in 1995, subsequently visiting bear country in various parts of Europe, Japan and Canada, seeing their footprints in the mud or in the snow, and hearing about how they live in the countryside in the Italian Alps and the Slovakian hills. They seem to survive alongside humans in countryside not unlike where I live in Scotland. There are problems, of course, probably the worst being when bears break into honeybee hives to eat bee grubs and honey, but in several places I was shown how this could be prevented quite easily by means of an electric fence. Brown bears can live on just about anything, alive or dead: vegetation, berries, acorns, small mammals, invertebrates and carrion. Unless people tame them on purpose with food, they are naturally shy, but when they become unafraid of people they raid rubbish bins for waste food. I remember being taken by Christoph Promberger to see the bears scavenging the big dustbins in the suburbs of Brasov, a large city in Romania. For some, these bears were a menace, for others a tourist attraction.

We might all like the idea of Hercules, and a teddy bear has been part of most our childhoods, but these are big, powerful animals that can be dangerous if injured by hunters or forestry workers. In Scotland we do have forests with enough food and areas large enough to hold a viable population, but I recognise that reintroduction would definitely be challenging from a human point of view. If other countries in the world are expected to live alongside lions, hippopotamuses and tigers, though, and value them highly for their beauty, their rarity and their tourism potential, then surely we could manage to do so with brown bears? As a starter, for the sake of the planet and for biodiversity, we could do what the group FAPAS are doing in Asturias – planting thousands and thousands of fruit and nut trees in the mountains, where small farms are abandoned and the land impoverished, to provide food for all creatures, including bears.

## Aurochs and traditional cattle

The mighty aurochs, or wild ox, was the origin of all domestic cattle. It was a massive animal, standing two metres at the shoulder, with long horns well over a metre spread. It was thought to be dark brown, with bulls much larger than the females. It was considered a fierce animal by early people, difficult to kill without harm, but probably excellent to eat. Julius Caesar wrote that 'they are a little below the elephant in size, their strength and speed are extraordinary, they spare neither man nor wild beast; they are killed in pits'. In Scotland, remains have been found in ancient settlements and in the peat; there's a skull in my local museum and the size is impressive. They were depicted in ancient art, and the spectacular paintings of aurochs in the

Lascaux caves in France allow you to imagine you are in at the heart of the hunt. I would love to see them. Ecologically, the aurochs was an important member of the fauna in the early millennia, after the Ice Age in Britain; its huge size and behaviour creating pastures, highways and glades. It was estimated that in the Mesolithic period there were about 84,000 aurochs in Britain: compare that to the nearly 10,000,000 domestic cattle in Britain in 2016. Aurochs were widely spread across the European landmass, with the last one dying in Poland in 1627.

Aurochs were domesticated in the Near East about 9,000 years ago. About 5,000 years ago the Bronze Age people, whose stone circles show so clearly in the landscape of Scotland, brought from the Mediterranean the Celtic shorthorn, a much smaller, dark, hairy cow. At that time, the wild ox would still

The ancient skull of an aurochs dug up in Moray, currently in Falconer Museum in Forres.

have lived here but, slowly, the human population pushed it into remoter areas, for it would have been a competitor with their own cattle, in fact it would have been a nuisance. Finally, at some point, they were gone.

The idea of bringing back the aurochs is simple, then, for while there are none to reintroduce, their ancestors – especially the older, hardier breeds like Highlanders – can still perform most of the ecological benefits of the extinct wild ox in the countryside. There have been attempts to recreate the aurochs, especially in the 1930s, when two brothers in Germany crossed a range of different cattle breeds to create a large cow called Heck cattle. I have stood near a great herd of them in the Netherlands, and seen a small number on a friend's farm in Devon, and they are truly impressive. Others, though, with the advantages of present-day DNA profiling, are trying to rebuild a better representation of the ancient ox. It will be interesting to see what they come up with.

From an environmental point of view I have always believed that some of our native hardy breeds of cattle, such as Highlanders and Galloways, could be used in many nature conservation areas to replicate the activities of the ancient ox. This would be especially true if they were bred back to be capable of surviving all year round on natural vegetation, and were not to receive supplementary feeding or medicines.

I've always been happy near cattle, for when I was young I sometimes milked by hand and helped to make butter and cheese with the churn; and in my middle years, when we had a herd of hill cattle, I loved to walk with them to new pastures on the hills. I knew all their names and most of them would let me give them a rub, for we had known each other since they were born. These animals lived outside all year, snug in the forest at night and not dosed with avermectins, so their steaming dung

A herd of awe-inspiring Heck cattle at Oostvardersplassen in the Netherlands, replicating the wild.

was perfect for invertebrates. In the 1990s I wrote a booklet for the Cairngorms Partnership, which set up the National Park, called *The Importance of Traditional Cattle for Woodland Biodiversity in the Scottish Highlands*.

I was becoming alarmed at the disappearance of the traditional crofters' cows, which for centuries had ranged the hills, forests and low-ground pastures and marshes, but which were vanishing as farms 'improved' their breeds in the New Agriculture. The arrival of massive continental cattle like Charolais, Limousin and Simmental brought dramatic change; they had to be kept on improved pastures and fed in the winter. In the 1970s the old traditional cattle, which worked with nature, were vanishing faster than snow off a dyke on a warm day. Those older cattle, with their ranging behaviour, kept glades and flushes open, maintained flower meadows and trimmed the willow scrub. Their heavy hooves pushed tree seeds into the soil and created a network of tracks of great value to other woodland creatures, such as family groups of young black grouse and capercaillie. Their activities led to patchiness in woodland cover and a richness of ecotones. The alternative, without cattle grazing, was widespread leggy vegetation, providing poor-quality habitat for flora and fauna. The older cattle's heavy size caused structural changes in plant communities, such as encouraging tree growth, and created open areas and disturbed water margins to the benefit of a range of smaller wildlife.

Traditional cattle are one of the big players in recycling plant material, for increasing plant biomass and for diversifying plant communities in our natural areas. Their dung is a valuable habitat colonised by an incredible number of invertebrates, providing an important food source for many birds and mammals; it enriches the soil with a rich supply of nutrients via urine and dung. One cow not only produces about four tons of dung per year, but also an annual insect population weighing about a quarter of her own body weight. I am told that because their diet was so varied and natural, the production of methane is less than that from cows reared in intensive systems. For me, nature conservation is about more than the protection of individual species or habitats. It is to do with life on earth. It is the optimum use of the sun's energy to create plant biomass, which in turn is eaten and digested by herbivores, and thus by carnivores.

At that time I was not only wanting to maintain the ecological benefits of traditional cattle: I was hoping to secure a better future for crofters and farmers in the hills and islands, who farmed in ways beneficial to nature. My document was well received by farmers, landowners and foresters in Scotland and elsewhere, but rather fell on deaf ears in the conservation bodies and the government. The numbers of traditional cattle continued to fall; there were no major increases in financial support for these cattle, while paperwork and regulations were most irksome for small producers. Yet the requirement ecologically to graze the greatly increased areas of nature conservation and rewilding was ever greater, especially when deer numbers in those special places were reduced and the heather grew longer. In 1998 I wrote that we badly needed to have more of Scotland in prime ecological condition. My premise was that it was impossible to restore our natural forest and moorland ecosystems without cattle, one of the most beneficially influential herbivores. Alongside beavers, elk and wild boar, they were key herbivores in the primeval Scottish forests. Nowadays, I recognise that this will not be done solely within traditional agriculture, where the finished beast is sold at market, even though there is a premium for such specially reared beef. I'm sure the future is in having traditional cattle, including rare breeds like Chillinghams, grazing nature conservation areas for nature conservation purposes. Those cattle will live their whole lives there, contributing to the biodiversity of our planet, and as the years go by will grow thicker coats for survival and lay on more fat for cold winters. Their intestines will lengthen again, reverting to how they were in olden times, able to deal with poor-quality vegetation in winter and early spring, and they will learn how to live efficiently within their home ranges. Although they will be much smaller, they will be restoring most of the attributes of their ancient aurochs ancestors, bringing the best of the past to help our planet's future.

## Wild boar

Wild boar were common and widespread in earlier millennia: research carried out by Derek Yalden at Manchester University estimated that there might have been nearly a million wild boar in Britain 7,000 years ago, when forest cover was most widespread. It remained a well-known animal into the start of the last millennium, with the hunting of wild boar depicted on Pictish standing stones and the animal remembered in place names. Wild boar were first domesticated over 10,000 years ago in the Fertile Crescent, in present-day Iraq, and – interestingly – wild boar in China were domesticated independently in about the same era. Early farmers brought them to Europe, and probably to Britain, in the Bronze Age. Early on, country people domesticated wild boar in Britain by rearing piglets taken from the wild, for this was an easy way to obtain meat and pig products, but progressively they bred with imported pigs; and the last wild boar, with their iconic, characteristic features, were lost with the advent of improved agriculture. The last in Scotland may have been several centuries ago. Farmers continued to herd domestic pigs in the countryside, however, especially during the autumn fall of acorns and nuts.

I remember groups of pigs rooting away under big oaks in the New Forest when I was a boy, and we kept a pig or two, for food, in the barn at home. When they were let out in the field, they could rapidly turn over the soil in their search for food. Wild boar are omnivorous and eat whatever is available, including carrion, and while some regard their rooting activities as highly damaging, others believe that it is ecologically beneficial for soil aeration and health. The creation of open areas encourages plant and tree regeneration, and there's evidence that pigs are beneficial for breaking up large swards of bracken. Concerns about potential damage to bluebell woods and other flowering plants would only be justified if pig numbers were allowed to become too great, for wild boar and bluebells have coexisted since they evolved. The only difference of course is that, for most of their existence, wolves and other predators exercised a natural control on pig numbers.

I first visited the Netherlands in the late 1960s, when my hosts tried to show me wild boar in forests near Arnhem. I was unlucky, although I did see where they had been rooting. Later visits to France, the Netherlands, Germany and Romania brought me in contact with wild boar, usually just fleeting views as they dashed off through the woods, but sometimes better ones at night, family groups caught in car headlights as they fed along the verge. I found it interesting to see a big animal that could make such an impact on the forest floor. In the 1970s there was recognition that rearing wild boar on farms in England was a way of making a good income from quality meat. They required special fencing to keep them from escaping, as well as licences to keep them, but the venture was successful enough to encourage other people to start wild boar farming too. But, as anyone who's ever kept pigs will know, they are masters at escaping. Some say that the great storm of 1987, which blew down 15 million trees across England, also brought down fences and allowed wild boar to escape and create a feral population in Kent and Sussex. Others escaped elsewhere in the country, and some were probably released for hunting purposes. By the end of the 20th century there were scattered populations living in the wild from the north of Scotland to the Channel coasts.

The first wild boar that I saw in Moray, where I live, was in November 2000, at a farm near Buckie. The farmer took me to look at three enclosures in the Scots pine forest; two held breeding groups of wild boar and the other had domestic pigs. He was hoping to make and sell Parma ham with a Scottish provenance. A year later, I looked at another enclosure of wild boar in a natural wood in Moray: there were 30 piglets with the sows and I thought they looked absolutely great, rooting around in such natural surroundings. Some years later some escaped and were seen living in the wild on various large country estates; I even saw where some had been eating mussels at low tide at Culbin Bar on the Moray coast. It reminded me of the wild boar eating freshwater mussels in big, dried-out lakes in northeast France. I also saw ecological experiments with wild boar at the Alladale estate in Sutherland, researching their impact on heather, and in Glen Affric, aimed at controlling bracken-dominated woods.

It's said by some that these wild boar are just hybrids, that they should be called feral pigs and exterminated when found

Wild boar in a plantation enclosure in Moray.

roaming free in the countryside. I think that would be wrong. They were here long ago and they still have an important role to play in the ecosystem, although we do need to learn how to manage their numbers effectively with their main predators missing. Furthermore, recent genetic investigations have shown that many free-living animals are clearly wild boar, rather than feral pigs. It's encouraging to reflect, then, that our wild boar are, in fact, probably as genetically pure as many of those we see rooting about in woodlands in mainland Europe.

## Elk

Most people are surprised to learn that the mighty elk – called the moose in North America – was once found throughout Scotland, England and Wales. The elk is a massive deer, with bulls standing over two metres high and nearly three metres long, weighing in at up to 800 kg. With broad, palmate antlers up to a metre in length, this is a very impressive animal. The females are smaller and lack antlers, and both sexes are grey-brown with paler legs. It's nowadays a mammal of the northern forests, in Europe ranging throughout Scandinavia and northwards from Poland and the Czech Republic. Worldwide, it extends across the boreal forests of Russia and North America.

Elk are herbivores, preferring in summer to browse on the leaves and twigs of shrubs and deciduous trees, especially willow and aspen, as well as plants in woods, meadows and marshes. In winter they browse on pine shoots and the twigs and bark of deciduous trees; nowadays, some also feed in hay meadows and in crops of cereals and oil seed rape. One of the most important ecological benefits of elk, in my view, is their ability to feed in marshes and freshwater, where they pull up and eat water plants, even submerging their heads to get to plants such as water lilies. They are essential for maintaining open water in forest bogs and are able to prevent small ponds and lakes, especially in woodland, being overwhelmed by vegetation. In many ways they complement the activities of beavers. In Scotland, forest bogs and pools are gradually grown over with matted vegetation and, in the end, disappear. On nature reserves, this succession – caused by the unnatural absence of large mammals – can be remedied by using mechanical dredgers and installing artificial dams, but that often needs to be redone every decade or so. Where I live, forest bogs and pools are very important habitats for birds such as teal, goldeneye duck and green sandpiper, a very rare wader in Scotland, and for dragonflies and amphibians. Elk also have the ability to sequentially graze bushes like

An elk walking and grazing through willow bushes in the snow with her calf.

willows and create a herbivory of fresh and nutritious shoots. Their huge body size creates pathways through bogs as well as through woodlands, beneficial to smaller birds and mammals living in the same habitats. These tracks are particularly useful to breeding birds with broods of newly hatched young, such as capercaillie, black grouse and woodcock, as they move between the best feeding places and safe shelters. It is hard to see how we can maintain the water element of natural forests without the big mammals like elk.

The elk was part of the Scottish fauna following the last Ice Age and ranged throughout the land. Mammal specialists at Manchester University estimated that in the Mesolithic period, 7,000 years ago, it was possible that the elk population in a wooded Britain could have been nearly 65,000 animals. There's much debate about how long they survived and why they became extinct. Carbon dating suggests an early exit, but elk remains were found in Roman settlements in the Scottish Borders and also in the 9th-century broch at Keiss in Caithness. It is thought it may even have lasted longer, with references to it in the Gaelic poem 'Bas Dhiarmaid', from the early 16th century, as well as the Gaelic name, Lon. It was probably exterminated in the early years of the last millennium through destruction of forest and human hunting. A full-grown elk would have provided a large amount of food for early humans.

In northern Europe, numbers were very low in the 18th and 19th centuries, due to heavy hunting pressure, especially in times of deprivation, but after the 1960s they boomed. In Sweden it was estimated that the annual harvest by hunters in 1960 was 30,000, but when I visited Sweden in the early 1980s it was 170,000, which suggested a population of half a million. Nowadays about 100,000 are hunted annually in Sweden and over 30,000 in Norway. In the past, some elk were domesticated for meat and milk, and very small numbers are domesticated in Sweden and Russia for meat, milk and even cheese.

I first got to know elk in the wild when I visited Sweden in 1979, and I got even better views of them when I returned to Värmland in 1983. The first thing that struck me was their great size: it was like seeing a horse walking through the edge of the woods. The latter visit was in July, and by that time the females were bringing their young with them when they were feeding at dusk outside the woods. On several evenings we drove along minor roads through the forests, past the scattered small farms. My diary records that on 16 July we had an evening meal of roast elk with our farmer friends, and then drove the local roads before dusk, finding a female elk grazing in an oat field on a local farm. With her was a superb youngster, gingery brown with big ears. A little further on there was another female, stepping out of the edge of the forest to get to better feeding, though I did not see her young, while in the next hay meadow was a female roe deer with three fawns. That evening I also saw about 10 families of red-backed shrikes – it was my best ever day for the species – while ospreys flew overhead and cranes called from the hidden mosses. By now I had got my eye in for elk and, one early morning, saw a female and young in a hay meadow and, a little further on, another female with young. When I walked in the forests and mosses I saw their signs: big tracks, easy to walk along, piles of distinctive large dung pellets and the occasional stand of young birches dying after being stripped of bark and twigs in the snowy winters. There I saw another value of the elk, too, for they had created dead wood for invertebrates and woodpeckers.

In Scotland, elk are kept at locations like the Highland Wildlife Park. Males kept in captivity can become unafraid of humans and dangerous, as I saw when I stood outside the elk enclosure at

Grimsö Wildlife Research Station in Sweden, a bull crashing the fence a metre away. I would not have liked to meet him outside the enclosure. In 2007 two elk from Sweden were released into a 200-hectare enclosure at the Alladale Wildlife Reserve in Sutherland. They had a single young four years later, but the habitat there was not really suitable for them to live in the wild. Reintroduction to suitable habitats, though, could well be successful. I remember talking with hunter friends in Norway and Sweden about present-day elk hunting, and heard stories of them quartering a shot beast so that the hunter and three friends could carry back a year's supply of elk venison for their deep freezers. This is another species killed out in Britain by humans because of their value in meat, hides and bone rather than for being a pest.

On other visits to the north I came across adult elk near roads and villages, often giving me a start as this massive animal loomed suddenly just metres from my car on a woodland road. On one occasion, while being driven by an osprey colleague in his car through the suburbs of a small town, we came across a yearling elk trotting down the road. My colleague told me that when the juveniles leave their mothers they can be a real traffic hazard, so we waited for it to exit the highway. I would love to see elk back in Scotland, but after meeting them on Swedish roads, I began to have doubts. Some time later, I read an illustrated advert for the Saab 9000 car and its excellent safety features, in which the company said, 'The adult elk is large, heavy and mostly dark and in winter the Swedish countryside is mostly dark as well. Which explains why surprised Swedish drivers and elk often collide. Fortunately Swedish roads are populated with many Saab 9000s, so drivers and elk survive.' It's nothing if not pragmatic. Should we in Scotland bear in mind the role and beneficial activities of this iconic species before entirely dismissing the possibility of reintroduction?

Golden eagle. (*Photo by Mike Crutch/A9Birds*)

7

# GAELDOM

## *Returning eagles to Ireland*

During the early 1990s, when we were in the throes of reintroducing the red kite to the Scottish Highlands, I met two Irishmen who were to become great friends. Lorcan O'Toole was employed by the RSPB to oversee the red kite project over several years, while Allan Mee carried out work as a seasonal warden at the RSPB reserve at Insh marshes. I enjoyed talking to them: they were really interested in raptors and we discussed the loss of most of the species from their own country. They were good craic and played shinty, despite the different 'caman', or stick, used in Scotland. They were great ambassadors for raptor conservation, and were brilliant at getting on with the locals in the Highlands, almost certainly helped along by their Gaelic heritage. Lorcan went on to champion the return of the golden eagle to Ireland and, with a couple of friends, set up the Golden Eagle Trust, while Allan went to California to work on condors and then returned to Kerry to reintroduce the sea eagle on behalf of the Trust. I am very happy that I was able to help both projects get started.

## Golden eagles to Donegal

Early in January 1987 Ciaran O'Keeffe, the Superintendent of the Glenveagh National Park in Donegal, wrote to me saying that he wished to explore the possibility of reintroducing golden eagles to the Irish Republic. He explained that the national park extended to 10,000 hectares of mountains and moorlands owned by the government. There were large numbers of sheep, as well as a substantial herd of red deer, with quite a lot of mountain hares, but that red grouse were scarce. Glenveagh had breeding golden eagles until 1915 but the gamekeepers on the estate exterminated them. There had been occasional visits of golden eagles from Northern Ireland and, in the summer of 1984, one stayed in northern Donegal for nearly a year. Ciaran thought it was an ideal place for a reintroduction and the park staff could carry out the work on the ground. I replied saying that it looked a very suitable place for golden eagles, but that I would worry about them when they travelled outside the national park. I asked if use of strychnine by sheep farmers might still be a problem. I would help, though, if I could, with the arrangements to obtain donor young in the Scottish Highlands. He replied saying that he would look in more detail into the use of poisons and said that the attitude of sheep farmers would be important. His head ranger suggested asking a 'real Scotch sheep farmer to talk about his experience', as it would be taken seriously by Donegal farmers.

I don't know where the years went, but on 1 June 1996 I visited the Glenveagh National Park with my wife. At the park headquarters we met Ciaran with his wife Kate and their children, and over coffee and cake we had a really good talk about the restoration of golden eagles, as well as a great deal of other news. We were taken up the glen that runs alongside the lough and then walked up through lovely woods onto the shoulder of a mountain. It was a great area of ancient oaks running up into scrub birch. The view across the lough was very beautiful, with the hills rising rapidly on the other side; it was very easy to imagine eagles soaring across this landscape.

We explored West Donegal, where some of the hills were newly covered in Sitka spruce plantations, just like parts of Scotland, but there were also large areas of heather moorland, some of it with natural regeneration of willow and birch. Near Ardara, I met John Hannigan, the ranger, and we looked at the habitats there, mainly of heathlands, before driving south to Killybegs. By 4 June, when we drove home, I had got a clear idea of Donegal and its potential for golden eagles.

A week later I sent a note of my findings to Ciaran, saying that I considered the national park an ideal release site for reintroducing golden eagles to Ireland. I thought it was big enough to support two breeding pairs and, if the birds ranged out to Errigal and Muckish Mountains in the north, Loughsalt Mountain in the northeast, Glendowan Mountains to the south and Crocknafarragh to the west, there would be room for three or four pairs. I also considered the country west of Killibegs, at Slievetoohey and Slieve League, to be good for eagles, as were the Blue Stack Mountains and the hills of Inshowen. There was certainly enough room and I also saw a great variety of potential nesting sites: small crags with vegetative ledges, big cliffs, gullies and large trees. I read the transects that Ciaran and colleagues carried out, and although it was not as good as the area near my home, it was better than the degraded landscapes of western Scotland. I noted that grazing pressure was declining and that there was good regeneration of both heather and deciduous bushes. I was also told about the sheep husbandry, whereby the lambs were not put back to the hill ground until they were older. Finally, I considered the woodland edge above Lough Veagh between the castle and Glenlack as a very suitable location to build release cages. It would give seclusion for the birds, plus superb perching and loafing areas post-release, close to a future feeding site.

In July 1996 my good friend Jeff Watson, a world-renowned expert on golden eagles, visited the Wicklows and Donegal with Jim and Phillipa Haine. Jeff had huge experience of Scottish eagles, gained during his long-term research in the Scottish Highlands. His report to the Irish Wild Bird Conservancy recommended that a proposed reintroduction of golden eagles to Ireland could be successful. He suggested that several pairs could nest in the Wicklow Mountains but that long-term viability there would be at risk, unless larger numbers were reintroduced and subsequently bred in Donegal. In his view, the future for the species in Ireland looked encouraging. He said that the suitable habitats in northwest Ireland were superior to much of western Scotland with regard to the prey available, and he felt that Glenveagh National Park offered an excellent location for the initial release programme. He was concerned about levels of poisoning and said that this needed attention, while also stressing the importance of a strong dialogue with the sheep farming community.

Later that year Lorcan O'Toole and Jim Haine asked me to visit Dublin for the weekend and give a talk at the Irish Raptor Study Group conference. It was great to learn that Jim and Lorcan were moving raptor projects ahead in Ireland, which meant my lecture on my experiences with reintroducing sea eagles, red kites and ospreys was very timely. The weekend was an excellent time to discuss a range of projects, and on the Sunday morning Jim and Lorcan took me to the home of Charles Haughey, the former Taoiseach, to bring him up to date with our various projects, especially the golden eagle proposal in Donegal. He was very excited and supported the idea; he also told us that a sea eagle had been seen over the Blasket islands in August. In the afternoon we had a very interesting drive round the Wicklow Mountains, to the south of Dublin, looking at the potential for raptor reintroductions. It was excellent-looking country for golden eagles, and there were plenty of big trees for eagles to build eyries. I was also told there were big numbers of deer, as well as rabbits and hares, but rather few red grouse. Later, 2007, these mountains were used as a release site for a successful reintroduction of red kites.

Lorcan and Jim compiled a document on behalf of the Irish Raptor Study Group, Birdwatch Ireland and the National Parks and Wildlife Service, Glenveagh National Park. The 'Reintroduction of the Golden Eagle to the Republic of Ireland' (in Gaelic 'An t-Iolar Fīrēan a Athbhunū Ēirinn') presented a well-researched proposal and said that there was 'unique opportunity to restore the epitome of Ireland's Natural Heritage, as we enter the new millennium'. Over the months, I had frequent conversations with Lorcan about his project and how we could help. His hope was that he would receive up to 15 golden eagle young each year for five years, and I said we would give assistance in identifying sites and talking with landowners to get access permissions to take young eagles. I agreed that we would also help with the legal side; I would apply for the necessary licence to take young golden eagles from the wild and would register buildings at my home as an agreed site under the Balai Directive, which would facilitate the holding and transfer of young eagles to Ireland. In November 1999 Liam Lysaght phoned from the Irish Heritage Council to discuss the golden eagle project, and I was able to answer various queries and give it my full support. Clearly it was heading in the right direction.

After a disappointing delay in 2000 caused by a research sub-group of Scottish Natural Heritage chaired by Michael Usher, Lorcan finally got permission to start and the necessary licences from SNH in 2001. The first eaglet that I helped him to collect was in Badenoch. We met at 11 am and drove to the keeper's house, where we met Dave McGibbon and his young assistant. After a good chat about things, we set off into the hills in the Land Rovers. It was a brisk walk uphill to the eyrie, which was long established and an easy walk in on the steep slope. We found two young, with a big difference in size, so I ringed the larger one and Lorcan collected the smaller. The eaglets were really well provisioned, for there were 23 mountain hares, nearly all of them juveniles, 11 red grouse and one stoat, much of it already eaten. Lorcan placed the smaller bird in his special wicker hamper, strapped it to his back and we walked back down to the car. I placed it in the pens at my home and gave it cut-up rabbit.

The following afternoon, in really good weather, I met Lorcan, John Marsh and Kevin Collins from Ireland at a meeting place in a forest further west in Badenoch. That day's eagle eyrie was a much longer walk over the hills, disturbing frogs and lizards in the heather as we went. Looking across to the cliff, we could see two young, so Lorcan and John walked over and Lorcan was lowered down to the eyrie. He collected the bigger chick and ringed the other; there were the remains of some grouse in the nest and also an uneaten black water vole – the water voles living in the hills of the Highlands are nearly all black, make a good meal for young eagles and are often caught. We put this bird with the eaglet at my home and fed them. I then went to the Hebrides while the Irish team finished the collection of six young golden eagles and headed back to Donegal by air from Aberdeen airport.

Lorcan O'Toole (Irish golden eagle reintroduction project) and author on cliffs above an eagle eyrie in Badenoch, in the Scottish highlands.

When the young eagles were ready for release, Lorcan asked me to come to Donegal to help fix on radios and check the birds for release. John Marsh collected me at Londonderry airport and took me down to Church Hill, where we met Lorcan and Ronan Hannigan. We all went off straight away to the release cages at Glenveagh. All six eaglets were looking really good and fit, and ready for release. That evening it was extremely midgey in the glen, so we decided we would make an early start in the morning. The two groups of release cages had been built on a nice flat shoulder of the hillside, high above the lough, with gnarled old oaks and birch trees sheltering them. They were similar to the design that we perfected at Loch Maree and each compartment could hold either two or three eaglets. The young eagles had a beautiful vista from the cages out across Lough

Beagh to the far hillside of rough grasses and grey rocks, fringed by deciduous trees along the water's edge. Away to our left was a beautiful curving beach of yellow sand at the end of the lake. There was much for the eaglets to see while they waited for the day of their release, and that's all part of the 'hefting' process. The birds learn that this – rather than Scotland – is now their home.

The next morning, the weather was less damp and we could get on with catching each eaglet. When they were young in their eyries they were relatively easy to hold, but now, with fully grown talons and strong bills, we had to be very careful to keep them under control as we fitted coloured PVC wing-tags, collected measurements and weights, and fitted light VHF radio transmitter back packs. All went extremely well, and as soon as we put the eaglets back in their cages, they settled down and started to preen their feathers. We did notice that some eagles were pecking at the radio aerials and bending them, so we thought that it would be good to have stronger aerials in future years. Unfortunately the rain continued in the afternoon and Lorcan decided that it was too wet to release them, so I returned to Scotland, and the eaglets were released the following day. I could feel the great excitement from Lorcan and his colleagues that their project to restore golden eagles to Ireland had started.

The following year's golden eagle project started on 7 January, when Lorcan, Ronan and John arrived to stay at my home in Strathspey; over dinner at Coylumbridge Hotel I caught up on all the news of the eagles in Donegal. The next morning we had a very good meeting at Scottish Natural Heritage HQ in Inverness with Dr Jeff Watson, Hugh Insley of the Forestry Commission and Colin Crooke from the RSPB. It was agreed that every effort should be made to get more eaglets in the summer. On 14 August Lorcan collected me from Belfast airport and during our drive I caught up on his news about all the eagles, not just the new ones collected in 2002, but also the surviving birds from 2001. It was starting to look good. We went straight to Glenveagh, where I saw that the young eagles were looking excellent, even the two smaller ones. The next morning we went to the cages with four rangers from the national park; soon we were catching and wing-tagging the eaglets, and I trained Kevin Collins in wing-tagging and attaching the radio transmitters, so that he could obtain the necessary endorsements to his bird ringing licence. The two younger ones were not as well advanced and I thought that collecting young from poor-quality home ranges in Scotland was not a good idea. We just needed a bigger selection of permitted nests to allow Lorcan to choose the best donor young.

In 2003, Lorcan arrived at my home on 30 June on another collecting trip. The next day he was in Blair Atholl asking for permission, and he went back there on 2 July to collect a young eagle. On that day, I went to Lagganbridge to the eyrie we visited in 2001; Kevin Lawlor did the climbing, aided by his assistant, and the BBC producer Moira Hickey came with us to record a programme for Radio 4. As we approached the nest cliff, the male eagle flew west, very high, chased by a diminutive male merlin, which provided a beautiful contrast in sizes. The female was on the cliff and flew off east, leaving two big young in the nest. Kevin climbed down to collect the larger young and ringed the other, before abseiling to the bottom of the cliff and walking back to us, where I put the eaglet in one of Lorcan's specially constructed wicker carrying baskets. We then had the long hike back to the cars.

Early one morning Lorcan and I drove in the Land Rover to meet the Balmoral estate ranger, Glynn Jones, who took us across the high hills and along some incredible zig-zag mountain

Young golden eagle collected under licence, in special wicker carrying box for use in the field.

tracks to an eyrie overlooking the Angus glens. It was a beautiful location; we walked over the hillside and looked down into the nest, to see one young in the eyrie and the other on a ledge. I held the rope and Lorcan climbed down to collect the larger female. There were lots of grouse in the nest but, surprisingly, the eaglet was rather thin. We headed back over the hills, said our thanks to Glynn and returned to Strathspey. John Easton, our usual veterinarian for raptors, came to check the birds for animal health export and I filled in the Balai Directive forms so that Ronan and Lorcan could leave for Donegal early next morning. With the young that Lorcan and John Marsh had taken back to Ireland in June, the total for the year was 11 – a much better cohort.

In early August I met Lorcan and John, and with the help of Angus and Damian we measured, weighed, fitted wing-tags and radio transmitters in the morning, and finished the other six young in the afternoon. There were more females than males and they all looked in very good condition except for one, which had a damaged elbow joint. The next morning we returned to the glen and put fresh food in the cages, before hiking through the forest looking at potential nest sites in cliffs and trees, as well as checking out where free-flying eagles were fed on carrion. We also drove right up to the pass where there was another future breeding site in the crags, before heading back.

By this time the annual visits from Lorcan and his colleagues were part of my summer calendar and it was great to be out in the hills with my Irish friends. But I knew from past experience that the real work of eagle reintroduction was being carried out by Lorcan in Donegal: the daily task of finding supplies of food for the eagles, cutting it up and delivering it to the cages and then, for the rest of the year, making certain that the carrion feeding sites, for the free-flying eagles, were maintained with regular supplies of venison, rabbits and roadkill. It's a tough and demanding job through all weathers, but essential for the success of any project.

In 2004 Lorcan and I went back to the easy-to-reach eyrie in Badenoch; as we walked from the road end with Dave, the keeper, there were hordes of mountain hares with lots of young, as well as several broods of red grouse. We saw both adults, and when Lorcan looked in the nest there were two five-week-old young, plus an unhatched egg – a clutch of three is unusual – as well as the part-eaten remains of nineteen hares, seven grouse and a weasel. We collected the larger of the well-fed eaglets.

Lorcan returned on another collecting visit at the end of June, and on 1 July we went to one of my favourite golden eagle eyries near Tomintoul, where I had ringed my very first eaglet 43 summers before. Colin Gibson, the keeper, drove us to the nest site in an estate Argocat, which saved a big walk. We scrambled down the heather banks to discover the female feeding two young. This beautiful eyrie was tucked in behind a rowan tree and needed just a short piece of rope for Lorcan to scramble into the nest; he ringed the smaller eaglet and we carried the larger one back to the vehicle. This batch of young was taken back to Ireland that day by Lorcan and Seamus. On 10 August I was with them in Donegal, and during the day we caught and measured, weighed, wing-tagged and fitted transmitters to nine eaglets in excellent order; the tenth, a smaller, weak bird, was

Well-stocked eyrie with two young and plenty of prey and a moulted eagle feather – highly suitable for taking a young for the Irish Reintroduction Project.

left for another day. The next day, before I went back to Scotland, we fed and checked the eaglets – nine were ready to go – and there was time for Lorcan to drive me round the national park and view the Blue Stack Mountains, as well as pick up radio transmissions of two of last year's young.

2005 was a bad year because Lorcan, after collecting several young, broke his wrist when walking back on a mountain track, near Kingussie, with an eagle in his pannier. His friend Tony, from Ireland, was with him and they had a very long, painful walk on a rocky track, through the Monadhliath Mountains, to get to their van. Tony called me on his mobile phone and I met them on the A9 road at Kinveachy to take the eagle, so that he could get Lorcan to hospital in Inverness. I fed the four eaglets next morning and then Tony drove them – and Lorcan – back home.

At the wing-tagging in August, Lorcan's arm was still weak, so Marc Ruddock, a Northern Ireland raptor worker, drove me from Belfast to Glenveagh. We processed all seven young eagles while I trained Marc and John in the techniques of wing-tagging and fitting transmitters. The excitement of this visit was when Lorcan took me to a line of cliffs where a pair of his eagles had built their first nest and laid an egg, which unfortunately failed to hatch. After a walk we sat in the heather and watched the female fly across and land in the rocks, then the male fly over the hilltop, and soon the pair flying together. I was very excited to witness another landmark in their project.

In 2006 I was presenting for a BBC film on seabirds in the UK, being shot that summer and, on 1 July, caught the early morning ferry from Mallaig to the island of Canna, via Rum, with the producer Roger Webb. We were there to film the exceptional long-term seabird work carried out by Bob Swann and his Canna team. When we landed, I found that Lorcan was already there and heading off to collect a golden eagle chick from an eyrie on the great north cliffs of the island. I joined him and watched as Justin Grant lowered himself on a rope down to the eyrie and collected one of the young. This is a beautiful site with huge, grassy slopes, rich with wild flowers, and broken up with steep cliffs, with huge numbers of seabirds on the water below. As we walked, with the eagle in the pannier on Lorcan's back, we stopped to watch a pair of sea eagles, and far below on the steep cliffs could see the two big young on their nest. For good measure, a local peregrine joined in to chase one of the sea eagles. It was a beautiful day in the Scottish islands.

I was disappointed that Scotland had failed to provide sufficient young eagles in the first five years of the project, which meant that Lorcan had to continue to visit us to get more. Most summers it was a struggle for him to get enough young, despite the help of the raptor groups and eagle ringers. Lorcan would range widely to get chicks, while others collected young and arranged for me to come and get them to add to the yearly cohort in one of my buildings. A keeper friend kept me well stocked with dead rabbits to feed them, before Lorcan would drive them to Donegal. It was always great fun to help Lorcan collect eagles, and I remember, in mid-June 2010, driving to Assynt in Sutherland to meet Doug Mainland and Derek Spenser and visit one of their nests. It was a grey day with low clouds as we hiked a couple of miles to a small cliff overlooking a remote loch. Doug and Lorcan went to the top of the cliff for Lorcan to use a rope down to the nest. He then lowered the two young to Derek and me at the base. We kept the bigger one for Ireland and I fitted a new-style GPS satellite transmitter to the other. The last collection for me was in 2012, when I went to Inverness to get a young one from the Uists from John Braine, an Inverness bird ringer. I looked after the eagle and fed it on rabbits. Lorcan arrived at my home on the 20th with another, and then set off for Donegal before dawn. The final tally over the 12 summers was 66 young golden eagles translocated from Scotland to Ireland, with 63 subsequently released.

The important action, of course, was still going on in Ireland. The project team were carefully monitoring the birds, often using the radio transmitters to track their movements, while sightings of the uniquely marked wing-tags gave data on their wanderings. As the eagles matured they started to settle down, and the first pair built a nest and laid eggs in their home range in 2005, while the first young was recorded in 2007. Occasional birds were lost to poisoning, while Lorcan also noted that in some areas of the degraded uplands the available wild prey was

The first golden eagle young reared in Donegal from reintroduced eagles – in the nest before fledging; note badger as prey. (*Photo by Lorcan O'Toole*)

too scarce to provide food for successful breeding. Between 2007 and 2018 twenty young eagles were reared. A big step forward was taken in 2018 when the first eaglet was successfully reared by Donegal-hatched adults. By that year, the eagles were still living in Glenveagh National Park but had also spread out to the Blue Stack Mountains, Glencolmcille and Inishowen, as well as into the glens of Antrim and County Down in Northern Ireland. It was estimated there were between 27 and 32 golden eagles in the wild at the end of 2017. It had been a hard-fought success made possible by the dedication and hard work of Lorcan O'Toole, his colleagues in the Golden Eagle Trust and the National Parks and Wildlife Services, and the local people in the northwest of the island of Ireland.

At Glenveagh Lorcan demonstrated clearly that the techniques for rearing and releasing translocated golden eagles, including the design and location of hacking cages, had been perfected. But the project revealed some of the problems associated with the reintroduction of large birds. The key is to release sufficient translocated birds, but just as important is the duration of the translocations. Of course, some of the choices are down to the logistics of collecting, caring for and releasing larger numbers, but it is clear that the larger the numbers and the shorter the period of time, the more successful the project. The different year-cohorts then have a greater choice when finding future mates of similar ages. In my view, releasing, say, 30 young per year over three years at three locations in a region is preferable to releasing 10 per year for nine years. Living in the donor location of the project, I remained disappointed that Scotland was unable to provide more eaglets. When I think about how the Swedes provided 93 young red kites in five years, or the Norwegians 100 white-tailed eagles for Ireland in five years, I still feel embarrassed.

Many potential donor sites were unavailable to the Irish project because of Special Protection Area status under EU legislation, and these often included the pairs breeding in the east of the country, which were more likely to have two strong young. Surely our government could have worked out a solution to this bureaucratic hurdle for the sake of such an important nature conservation project? To be honest, the ban against the taking of solitary young from productive pairs is more to do with social attitudes than ecology. Opposition from the gamekeepers' organisation, which had been criticised for the illegal killing of eagles, led to the government banning the taking of young from the areas east of the Great Glen in 2007, and this was unjust. It impacted on the project through spurious thinking, which involved most of the nests with two young. But, despite all that, the golden eagle has been restored to Ireland, which has strengthened its conservation status in Western Europe, and it stands as a beacon of hope in a damaged natural world.

## Sea eagles to Killarney

20 May 1985 was one of the most bizarre days of my life. I caught the early morning Otter aircraft down from Inverness to Glasgow and then an Aer Lingus flight to Dublin, where I was met by Richard Nairn, Director of the Irish Wildbird Conservancy, who had invited me to Ireland. We were on our way to have lunch and a discussion with the former Irish Taoiseach, Charlie Haughey, at his home outside Dublin. Mr Haughey was very interested in sea eagles and wanted to reintroduce them to Ireland, in fact specifically to his own island of Inishvickillane in County Kerry, one of the Blasket islands. I was there to give advice. He and his wife Maureen warmly welcomed

me and introduced me to the lunch party that included the ornithologist David Cabot, wildlife filmmaker Éamon de Buitléar, explorer Fergus O'Gorman, friends from the government and a Monsignor of the Catholic Church.

We talked about sea eagles over lunch and I showed a film and slides of our sea eagle projects in Scotland. There were lots of questions and a wide-ranging discussion, and it was clear to me that Charlie Haughey was really keen to start as soon as possible, while others were doubtful. He was a keen environmentalist and during his time in government initiated various nature conservation gains, including a large protected ocean zone for whales. He had also moved a small herd of red deer from the mainland of Ireland to his island to prevent hybridisation with Sika deer. Before departing, we walked through a near-hidden side door of his lounge to enter a small and amazing replica country pub, with Guinness on tap, to have drinks and further chat. I remember him asking me if I thought restoring sea eagles was possible, and I said, 'Yes, but it will need a lot of eaglets from Norway and really good funding.' His reply was, 'I thought so – for most things can be sorted with a fistful of fivers!'

Following the meeting, Richard prepared a confidential report on the reintroduction of white-tailed eagles to Ireland. I gave him help in drafting it and advice on contacts in Norway and Scotland. His final report was adopted at the Irish Wildbird Conservancy Council meeting on 15 June. Later, Richard visited the island of Rum and also corresponded with the Norwegians, with both Johan Willgohs and Harald Misund offering support.

In Ireland, though, he was getting a cautious response from the Department of Fisheries and Forestry, who wanted to make sure that the factors leading to the species' extinction in Ireland were no longer an issue. These concerns led to a delay and to a more detailed feasibility study of southwest Ireland. Ultimately, this revealed a major threat, at that time, to any future reintroduced sea eagles from poisoning, because sheep farmers were legally allowed to use strychnine-poisoned baits to kill foxes and stray dogs as well as alphachloralose-poisoned eggs to kill crows and ravens. On 10 October 1986 the project researcher Tony Whilde wrote to me with his findings – that large amounts of strychnine were being used – and asked my advice. I was horrified at the number of permits and the quantities of poisons that he had discovered were being bought from chemists by sheep farmers. I replied: 'I confirm that in view of the very distressing situation over strychnine use it would be unwise to release sea eagles in the present circumstances. This is really sad as it would be so great to have sea eagles and golden eagles back in Ireland.' A downhearted Richard wrote to me at the end of the year to report that the sea eagle project was back on the shelf. What a dreadful shame.

A year later, in November 1986, a large eagle did appear in County Kerry, and the wildlife service rangers asked me if it might be a young white-tailed eagle from Scotland. After seeing the photograph, I suggested that an expert ornithologist check it out because I thought it was a young bald eagle from North America. That is what it proved to be – an amazing stray, blown across the Atlantic Ocean. Pat O'Connell, the ranger who nursed it back to health from its exhausted state, accompanied it back to the United States courtesy of Aer Lingus. Charlie Haughey, now once more the Irish Taoiseach, said as it departed: 'Iolar, I wish Godspeed to our feathered friend. May he live long and happily in the wild back in his natural habitat.' In 1992 Charlie finally got his wish when two white-tailed eagles from the Berlin Raptor Centre were released on Inishvickallane; soon after, one was found dead near Tralee, while the other may have survived longer.

* * *

In 2007, after several years of planning and building on the earlier proposal, the Golden Eagle Trust and the Irish National Parks and Wildlife Service started a full-scale reintroduction in the Kerry National Park. Sadly, Charlie Haughey had died the previous year so did not witness the start of the successful project. A very strong feasibility report had been put together in support of it by the three Norwegian sea eagle experts, Duncan Halley, Torgeir Nygård and Alv Ottar Folkestad, and they agreed to provide 100 young from Norway. Allan Mee had come back to Ireland, after his years with the Californian condors, to run the project on behalf of the Trust. It was a pleasure to hear from Allan again when he emailed me on 11 January asking for my help in wing-tagging and fitting the radio transmitters on the young sea eagles when they were ready to be released in late summer. I said that I would be delighted to come over and help and that I hoped they would get a good number of young in their first year.

In July we made arrangements to go to County Kerry, and on the 22nd I flew from Inverness to Dublin and on to Kerry airport, where Allan collected me and took me to his home. The next morning we met Lorcan, Barry and the veterinarians at Killarney National Park headquarters. We picked up more people on the way round the lake to the release site; Allan and his colleagues had constructed the release cages in a wooded hillside overlooking the lake. They were a very well-made series of cages, similar to the ones used in Donegal for the golden eagles and the Wester Ross sea eagle cages. The view for the sea eagles, and for us, was spectacular, looking down over the mixed conifer woods below to the wooded islands, large and small, in Lough Leanne with, in the distance, the town of Killarney.

We worked all day, catching up the young eagles, two or three in each cage; the veterinarians examined them for health assessments and took samples of blood for analysis and DNA determination, while we weighed and measured them, and fitted coloured PVC wing-tags and harness-mounted satellite transmitters. We completed 12 of the 15 young ready for release, while the three smaller ones would be processed later. We returned to Allan's home to eat and discuss the day; the transmitters were all working fine when checked by computer.

Next day we went to the national park workshops to collect food for the sea eagles and then set off for the release site. All the eagles were looking extremely well after yesterday's activities, and when I looked through the peepholes I just loved to see how these large raptors seemed to enjoy perching side by side on the branches. Sometimes they would even pick at loose down

Young sea eagles wing-tagged and fitted with transmitters – ready for release in Killarney, Ireland, July 2007.

on each other's plumage, surprisingly gently for such huge eagles. We pressed on with the remaining three birds and fitted transmitters to two of them. We drove the long way back so that we could view the cages from the far hillside; from there, we decided that some of the trees needed pruning to provide perches and safer flight routes. Before leaving for Scotland I had a day exploring the Dingle Peninsula with great views to the Blasket islands, including Inishvickilane, and finally drove back to Allan's house via the Gap of Dunloe. All of it looked excellent white-tailed eagle habitat and I felt sure that their proposal would succeed.

Allan asked me to come back in July 2008, when they had collected 20 birds from Norway. As before, he collected me from the airport and we drove to his home in the Black Valley. In the evening we had supper at Kate Kearney's with Damien and Ann Clarke from Wicklow, giving us a great chance to catch up on all the raptor news in Ireland, including the exciting news of a red kite reintroduction in the Wicklow Mountains. The next day was a complete wash-out as it absolutely lashed with heavy rain all day. In fact, we did not leave the house until there was a slight gap in the rain in the evening. We went up to the cages to check out eaglets before dark and they were all looking sodden, while the streams in the forest were raging torrents. The 30th, luckily, was a warm sunny day, though, ideal for our work. We reached the eagle cages at 7 am and started catching up the eaglets half an hour later. Allan and I worked one cage at a time and had completed four birds by the time Neil and Pat arrived. Then we really got going with weighing and measuring, and fitting the transmitters and wing-tags. By 3.30 pm we had used all the satellite transmitters on 16 eaglets, all of them in excellent condition. The other four were left for later. During the day there was also a chance for me to add my full support for Allan's application for wing-tagging and transmitter-fitting endorsements to his ringing licence, which I was pleased to do. Allan fed the birds before we left the site, and that evening, after a clean-up at Allan's house, we had a long yarn over a bar supper. It had been a privilege to be asked to help at the start of the Irish project, and from then on I followed their reintroduction of white-tailed eagles with keen interest.

The successful project went on to release a total of 100 young white-tailed eagles, taken under licence from nests in Norway, in the Killarney National Park in southwest Ireland. The team carried this out in five years from 2007 to 2011. It involved a massive fieldwork effort by their Norwegian colleagues, who succeeded in collecting a remarkable number of young eagles. There was also the huge task of feeding and caring for them in captivity, carried out by Allan Mee and his colleagues. In addition, there was massive interest from the public, and the team worked hard on meeting farmers and local people wherever the sea eagles ventured. The first pair formed in 2010, quickly increasing to four pairs in 2011, ten pairs in 2013 and 14 pairs in 2014. The first pair nested in 2012 in County Clare, and the following year this pair reared the first two young sea eagles in Ireland – a momentous occasion. Since 2013 a total of 25 Irish-bred eaglets have fledged in the wild. In 2018 there were ten breeding pairs and the team were encouraged with the progress of the population to date. There have been losses to illegal poisoning, to disease and to full-grown individuals fighting to the death. In 2006 and 2007 there was concern on the part of sheep farmers, and lobbying of the Irish government, but since then there has been no proven case of an eagle taking a lamb, even where pairs are breeding in hill sheep areas. The sea eagles are now seen as very much a part of the landscape of Ireland and I am proud to have played a part in their story.

Male osprey turning in flight, ready to dive into the loch to catch trout. (*Photo by Mike Crutch/A9Birds*)

# 8

# GLOBAL VISION

## *Restoring ospreys in Europe*

In 1996 the American-based Raptor Research Foundation held an international meeting on raptors in Italy. Locally it was organised by Massimo Pandolfi in Urbino and we had an osprey symposium on 5 October, at which many of my osprey friends were giving presentations on mainland European ospreys. My presentation was not only about the conservation of – and increase in – the species in Scotland, but also looked at the translocation of eight young in the summer from Scotland to Rutland Water in England, the start of a five-year project. It was the first for ospreys in Europe, so there was terrific interest, especially from the southern European countries with few or no longer any pairs. During any spare time through the five-day conference, the European osprey researchers often got together to discuss new projects, novel ways of catching and studying their favourite birds and also the merits or otherwise of translocations in Europe.

## Portugal

Luis Palma, an experienced raptor biologist from the Algarve, sought me out several times to discuss ways of saving tiny populations of ospreys, for in Portugal they were down to just one pair on the Atlantic coast. After the conference we kept in touch and this led to a suggestion from Luis that I should come to Portugal and talk their ideas through. A year or so later, on 26 September 1997, I was being collected from Lisbon airport by his colleague Pedro Beja. He drove me all the way south to our hotel at Milfontes, on the Alentejo coast. There we met up with Luis for a discussion and a lovely evening meal overlooking the Atlantic. The next morning I was up early to have a look at the estuary which was full of grey mullet, the osprey's favoured fish. After our breakfast, which I soon learned meant, for my Portuguese friends, the tiniest and strongest cup of coffee possible, which took the longest possible time to drink, we were off exploring.

We had a brilliant morning looking at the coastline all the way south, every now and then parking the Land Rover and walking to the protected coastal strip of the Costa Vicente Natural Park. The cliffs were not high but they were very broken and sometimes featured pinnacles holding white stork nests. I was shown an old osprey nest site near the lighthouse at Cape Sardao, which had been abandoned after the keepers shot one of the birds. The coastal scenery was superb. Near Alezar, we met the director of the Faro National Park. Further south were higher rocky headlands and offshore a dramatic sea pinnacle with an osprey nest on the top. It was the nesting place of the final pair of ospreys in Portugal. Earlier in the spring the female had become entangled in monofilament fishing net carried to the nest as decoration and had died, despite their having got the military to winch down a rescuer from a helicopter. It was very sad to think that the solitary male was the end of a line. We finally reached our small *pensión*, where we spent the night. We ate at a small rural restaurant where Luis Fonseca, the toothless owner and chef, cooked us a fantastic sea fish dinner, with excellent wine, finished off with local strawberry-tree liquor.

On the Sunday we were out early again, to high cliffs near the old coastguard station at Casteljo, which they considered one of the best places to release young translocated ospreys. I thought so as well, but suggested the cages should be on the top of the cliff, with artificial nests close by, rather than at the bottom. The

This osprey nesting stack off the southwest coast of Portugal was the last site used by the original population, and was re-occupied by nesting ospreys in 2014.

old building could be done up and used by the wardens. We drove on to the southwest corner at Cape St Vincent to watch Cory's shearwaters passing by, and then east to a stretch of south-facing coast, where Luis knew of two old osprey nests. Then we drove inland to his study area for Bonelli's eagles, with massive eucalyptus trees, pine forests and deserted farms. We saw several pairs of Bonelli's, one beside their nest in a high eucalyptus. We stopped for more strawberry-tree liquor and coffee from an old friend, and then drove north. The following day Pedro drove me back to Lisbon airport, past lots of white stork nests built on electricity pylons; in one place I counted a hundred. It was the first time I had recognised that storks could compete with ospreys for nest sites.

Following my visit, Luis sent me their thoughts on how to restore ospreys to the southwest coast of Portugal, and this was the basis of their feasibility report. The following summer Pedro and his colleague, Rita Alcazar, came to stay with me to learn about osprey projects in Scotland, having visited our osprey translocation at Rutland Water. On 29 July I showed them ospreys in my study area along the River Spey, with a visit to the famous Loch Garten reserve. The next day we went to the white-tailed eagle release site in Wester Ross, where I showed them the young eagles and the hacking cages; the peep holes in the back wall allowed them to view the eaglets, which were looking really excellent. It was good for them to experience our release procedures and talk with the warden. We drove further north in the hope of seeing one of the adult sea eagles at Gruinard Bay, but had no luck.

The two days gave us plenty of time to talk about translocating young ospreys and how a similar exercise could be carried out in Portugal. After dinner at my home I put them on the night train to London.

In late October 1998 I was back in Lisbon meeting Pedro and waiting for Jean-Claude Thibault to arrive from Corsica. He was an old osprey friend who was spearheading their recovery on that island. It was great to see him again and hear how the ospreys were doing. After a midnight arrival in Faro, we met the whole team in the morning, including Luis, Pedro, Rita and Jorge Safara from Portugal, Eva Casado from Seville, Jordi Muntaner from Mallorca and Pertti Saurola from Finland. In fact, we needed four cars to get us to the western Algarve to view potential osprey release sites on the south coast and then, after visiting Cape St Vincent, we travelled north to their best potential osprey release site. The Natural Park staff had rebuilt the hut which looked over the cliff at the potential release sites. We ate lunch at the raptor watch point and watched 11 short-toed eagles soaring overhead, as well as a Bonelli's eagle being mobbed by a hen harrier. Later we all had a lovely congenial evening in the windmill restaurant at Bispo, talking ospreys and eating fine local food. The next morning we set off to look at locations on the Alentejo Coast, north as far as Cape Sardao, where two journalists got interviews about the osprey project. We visited the proposed hacking facilities at Torre de Aspa, and met with the Mayor of Vila do Bispo county. The old osprey nest on the cliffs had been taken over by white storks, and I thought again how that species would prove a difficult competitor for recovering ospreys on the sea cliffs, because the stork population had risen so rapidly.

We drove north to Odemira and so to the headquarters of the Natural Park, where our afternoon osprey meeting was held in the lecture theatre. Luis introduced the project and what had been learned, and I followed with a presentation on the Rutland Water project and my thoughts on Portugal. Jean-Claude gave a presentation on Corsican ospreys showing how, when the population increased, the breeding productivity slowed down, which was a pity, as Luis was keen to obtain young from there. Andreas Helbig from Germany discussed the DNA of ospreys and considered that the difference between northern and southern ospreys was not an issue. We had a really good debate and made a decision to go ahead, initially with Finnish birds as well as some young from Corsica and even, possibly, the Balearic Islands. I floated the idea of a recovery plan for the western Mediterranean. I also had an opportunity to talk with Eva about her survey of wintering ospreys in Andalusia, and I could immediately see that she and Miguel Ferrer, her manager at the Coto Doñana Biological Station in Seville, were exploring the idea of running a similar project in Spain.

Next day we drove to Lisbon for a crunch meeting at the Ministry of the Environment, where a small group of us met a senior official in the government and two research scientists. Luis outlined the project and discussed their feasibility report. I gave the latest details on the English project and there was a good discussion following the talks. It seemed to go well, but I'm afraid we later learned that the civil servants would not make a decision and give him and his team permission to start an osprey reintroduction to Portugal. It was a disappointing bit of news to hear from Luis.

But Luis did not stop dreaming about returning the osprey to his country. In the meantime he carried out research with colleagues on the ospreys of the Cape Verdes, and then – in 2010 – he made a breakthrough, obtaining a promise of funding for a reintroduction in the east of Portugal. He formed a partnership with Finland and Sweden to send young ospreys to his new facility, and they were able to start in 2011. He had asked me for support and advice, and although I was unable to collect any young ospreys from Scotland, as we had promised to support a new project in the Basque Country, I was still supportive and ready to help as an adviser.

On 7 March 2012 I was collected from the airport by Andreia Dias and Pertti, and Peter Lindberg arrived soon after from Stockholm. We drove across country to Evora and then by ever more minor roads to a small hotel near the massive reservoir of Alqueva, where we met Luis and Eva. The next morning we were at the edge of the lake admiring the tiny cottage they had renovated and fitted with solar electricity and piped water. It was a beautiful, hidden locality surrounded by cork oaks, with hoopoes and black-shouldered kites to gladden my northern eyes. The release cages that they had used the previous year were several hundred metres across the peninsula, facing out over the water. They were elevated off the ground, as there were both foxes and wild boars in the area. We had a good search through the surrounding area to look for good perching places and potential nest sites. My first impression was that, when the birds started to nest, it might be difficult to find them because there were so many potential nesting sites on this huge and convoluted reservoir, 83 kilometres in length, with multiple inlets and about 200 islands. The lake was full of fish such as carp, goldfish and bleak. We had an alfresco lunch in the sunshine, when we discussed the project, including the loss last summer of a bird's tail feathers which had held the tail-mounted transmitter. We talked on through the orange-red sunset and then by the light of a brilliant full moon.

Next day Luis had organised our meeting in the wine cellar of a small castle. In the presentation room upstairs Luis gave his report for the previous year and opened a discussion on the translocation, care and release of the ospreys as well as subjects such as veterinarians, volunteers and business partners. It was an encouraging and interesting meeting, followed by a visit to the extensive wine cellars. We travelled to look at the reservoir dam, and ate our lunch with a view over it, before Eva took Pertti and me in her car to drive to Seville to discuss their project.

Two years later, in the autumn of 2014, I returned to Lisbon to be met by Luis for a tour of the city. Once Björn Helander, Peter and Pertti had arrived, we began the long drive east to our *pensión* at the Alqueva Dam. The next morning, Joao arrived in the team's boat to take us to the release site. Before landing, he checked the nets, which had caught some fresh barbel for the ospreys. We walked over to look at the release cages, which were looking good again, then went on up a small hill to overview the area, as five of the young ospreys had not yet migrated. We saw two of them perched further along the shore. When we

Hacking cages with four compartments for releasing translocated young ospreys at Alqueva Reservoir in Portugal.

got back to the cottage, Andreia had arrived with fresh sardines from the coast and we had a great barbecue in the sun, ideal for a leisurely discussion of progress during the year. Later, after the osprey team had put out fresh fish at the cages, we watched as all five young ospreys flew to the platforms to feed. It was excellent to see.

On 12 September we again went by boat to the field base and had a far-ranging discussion. Andreia told us that three young ospreys had departed early that morning. After midday we set off for the Algarve, and I was extremely interested when Luis pointed out a golden eagle nest in an olive plantation, at which the pair had reared three young within sight of the local road. That's how some pairs should live in the UK, rather than remain banished to the uplands. Luis then gave me another brilliant surprise: a privileged visit to the Iberian lynx breeding facility for he knew that I dreamed of seeing lynx back in Scotland. The next morning we were driven to the top of a 1,500-metre mountain for fantastic views across southwest Portugal to the Atlantic Ocean. We checked out several reservoirs as we headed for the famous last nest on the Alentejo coast. A check with the scope showed it had been recently built up and used by ospreys, which was confirmed by Luis in 2015, when a pair reared young. The team went on to complete the release of 56 young ospreys, 33 of them males, and their first nesting in 2015. In 2018 there were three pairs at Aiqueva, one pair back on the coastal pinnacle and one pair on the estuary of the River Guadinana. Luis and his team, after an initial setback, had restored the osprey to their country.

## Andalusia, Spain

Following Eva Casado's talk about osprey reintroductions in 1998, she and Miguel Ferrer had been in contact with me over the possibilities of running a project, and on 14 March 2002 we finally had the chance for a very thorough discussion on the whole subject in Miguel's office in Seville. He decided that he was ready to start a reintroduction, and Eva's fieldwork on wintering osprey suggested that the reservoirs in southern Andalusia and the coastal estuary at Odiel might be the best places. During the previous few days I had taken a really good look at some of the possible sites. At the Odiel marshes, west of Huelva, the Natural Park director, Juan Rubio, and warden José Sayago showed me some superb coastal habitats, where we saw six ospreys; José told me that about 20 wintered, including three from Scotland. One day, we went by rubber dinghy to a salt marsh where a bird he wanted to catch had been perching on a pole. He swapped the pole for one with his osprey snare and he hid me under camouflage netting, and they went away. Half an hour later his voice came on the radio calling 'Roy! Roy!' and I could hear the bird above me and then immediately it was caught. I ran to the trap, extracted the bird and got a real surprise when I saw she was carrying a Scottish ring – she had been ringed at Glentress Forest in the Scottish Borders.

Later, Eva collected me at the Biological Station in Seville and we drove down to Cadiz province to check out some very good osprey habitats and saw seven ospreys. The next day we visited the reservoirs of Barbate, Guadalcacin and Bornos, and later the salt pans at San Lucar, meeting local osprey enthusiasts on our way. The last day in the field started with me giving a talk on osprey reintroduction at the Botanic Gardens at San Fernando, before exploring the coastal salt pans near Cadiz and north through the marshlands west of the River Guadalquivir. The last ospreys had nested in the cliffs south of Cadiz 70 years before, but I had no doubt that they could be restored as a breeding species. Later in the year Miguel came to my home in Scotland and I gave him a tour of osprey nesting sites in several of my study areas; and, for the record, we visited the famous tree at Loch Garten. We made plans for the Spanish project while driving in the Highlands and continued at home.

In March 2003 I returned to Seville, and in the evening, with Pertti and Eva, talked ospreys at a bar and watched the lesser kestrels on the cathedral. The next morning we headed south via Lago Medina – five white-headed ducks were a bonus – to Alcala los Gazulos. We met local people, including the environment staff and a landowner, at nearby Barbate reservoir. The farmer guided us to various parts of the reservoir and I then found what I thought would be an excellent release site, among scattered oaks on a slope running down to the water's edge. In fact, it was the site the project went on to use, right from the very beginning. We stayed for three nights at the old convent in the hilltop village of Vejer, for an osprey conference. Before then, Miguel and Eva took me to meet a landowner at the Guadalcacin reservoir, where we were shown nesting sites of Bonelli's eagles. Here, again, there were abandoned small electricity pylons standing in the reservoir, which struck me as ideal nesting sites for ospreys.

The conference started in the evening and continued into the next day with a whole series of talks about ospreys in southern Spain and their proposal to start an osprey reintroduction. Pertti and I both gave lectures. The thrust of my talk was that the osprey was a pan-European breeding species which had suffered intense persecution and historical range reduction, and although northern and some central European populations were generally healthy, there was little or no evidence of major range recovery elsewhere. I recommended that a programme of reintroductions should take place in southern and western Europe in order to restore the species range. With success, this could nearly double the pan-European range and population size, and increase the long-term viability of the osprey. There was much discussion during and after the sessions among the 100 participants. It was great fun to meet so many Spanish ornithologists who were working on a range of different raptor species as well as studying the great migrations between Africa and Europe at Tarifa. Most people were really excited about the possibility of the osprey returning to breed again in Andalusia. The NGO Birdlife Spain was opposed to Miguel's project, but this did not delay it.

Back in Scotland I started the process of obtaining a licence to collect young ospreys for Spain, and finally Scottish Natural Heritage licensed me to collect and export five young a year for five years, starting in 2004. I next visited the project on 31 July 2004, a day when the temperature was 45 °C. We quickly drove to Barbate reservoir, where Miguel took us to the hacking site. Four young ospreys had arrived earlier from Finland and I met the hacking team; they were worried that the birds were too hot, so I recommended fitting a water sprinkler in the roof above the nest to cool them occasionally. After going with Miguel to see their imperial eagle translocation some kilometres away, I returned to the hack site. Mara was extracting fish from a net in the bay, which she and Mathias cut up to feed to the young. We chatted about feeding, rearing and releasing techniques and then, after they had put fresh fish in the cages, through one-way glass we watched the young feeding. I spent the next day with the hacking team, getting a better understanding of the potential feeding sites for the released birds and future nest sites. The young ospreys were looking in excellent condition, despite the heat. Eva drove me north the following day, via the Sanlucar salt pans and marshes, to catch my flight home. It was wonderful to see the project started.

In mid-January we had the first of what would turn out to be annual meetings of the osprey group at the Huerte Grande outdoor centre, with its cabins in the pines looking out to Africa. This first meeting was much more international than most, with osprey workers including Rolf Wahl from France, Mikael Hake from Sweden, Daniel Schmidt from Germany, Rafel Triay from Menorca, Pertti from Finland, and the Spanish team. Most of us were old friends through ospreys, so we had a wonderful range of talks and discussions, as well as hearing updates on the season and plans for 2005. On a field trip next day, we viewed the translocated young imperial eagle as well as taking a good look at the Barbate osprey hacking site, and the day ended with 1,200 cranes flighting at dusk to roost at La Janda.

In July I returned to Seville, this time with five young ospreys from the Scottish Highlands: they were TV news at both ends of the journey. Eva and Mathias drove me to Barbate reservoir where we placed the young in the hacking cages. The young from Germany were already there in other compartments and they looked very well. In the morning I netted fish with Pepe at the reservoir and then fed the ospreys. On Thursday Miguel drove me to the Odiel marshes to see the other release site. It was excellent, tucked against some trees, looking out over the salt marshes and creeks. Later, a project worker Roberto arrived with young from Finland to join those already there from Germany, as an adult migrant female osprey sat perched on an old electricity pylon in the marsh. The next day I headed home to Scotland.

Feeding platform for young reintroduced ospreys at Odiel marshes in Andalucia.

At my end of the project, I would work out which chicks were suitable for Spain as I carried out my annual monitoring. When we were ringing and colour ringing young ospreys in the first two weeks of July, I would collect five suitable young ospreys for translocation, which we then sent by plane to Seville via London. In total, 25 young over five years went to join 20 young from Finland and about 90 from north Germany. In 2007 Eva came to stay with us in Moray so that she could see the nests. Early each year, at the Spanish end, Miguel and Eva would organise the annual meeting at Huerte Grande or sometimes at the *estancia* at the Coto Doñana National Park, a superb place to stay with lots of local wildlife interest, especially a high viewing tower which gave a great outlook over the marshes. Our last flight with chicks for Spain was on 9 July 2009, when we sent off five young, with the local press and the general manager of Flybe watching. (We had, in fact, called one of the young Flybe.)

The first ospreys were not expected to breed until they were three years old, in 2006, but when we arrived at Huerte Grande in March 2005, Eva had amazing news: a pair of unringed ospreys were breeding. On 16 March we all drove to the Guadalcacin reservoir and were shown the nest on a small, unused electricity pylon standing in the water in a bay on the north side. With telescopes we saw that the female was incubating: she must have just laid her first egg. Sadly, some time later, strong winds blew out the nest; fortunately the pair built a new nest on the next pylon in the water and re-laid. The eggs did not hatch, and as a test, Miguel and his team put a young black kite in the nest which the ospreys accepted and fed, so two young ospreys were sent down by plane from Germany. The kite was placed back in its own nest and the two ospreys were reared successfully. The following summer, the pair nested once more but again failed to hatch their eggs, so two more young ospreys were substituted. Maybe they were very old ospreys, past breeding age, and – to judge by the choice of pylon nest site – probably from Germany.

In 2007 the first five translocated ospreys returned from Africa: three two-year-olds originally from Germany and two from Scotland. In 2009 the first successful breeding by translocated ospreys involved a female from Scotland with a German mate; they reared three young at Odiel marshes. The previous year they had paired up and built a nest on an unused electricity pylon in the marsh, but did not lay eggs. In 2010 they reared one young, and two in 2011.

By 2009 a total of 123 ospreys had been translocated from Germany, Scotland and Finland. The population grew slowly, but by 2015 17 pairs were breeding in Andalusia in the two areas in which they had been released. In fact, breeding numbers encouraged Miguel Ferrer to explore the potential of moving some young to Valencia, where I joined him in 2018 to talk with the local authorities and examine potential hacking sites and future breeding habitats.

## Catalonia, Spain

Manuel Pomarol, a biologist with the Wildlife Service of the Catalan Government, wrote to me in January 2002 about the possibility of reintroducing ospreys to Catalonia. He had written a draft proposal a few years previously and now his department wished to learn more. I invited him to come and stay with me in Scotland to discuss it and learn our techniques. I collected him from Inverness airport at midday on 29 July and on the way home showed him the famous osprey nest at Loch Garten. Later, we visited three other osprey nests in my study area, before discussions over a meal at my house. The next day we went to monitor a range of nests, with flying young, in Moray, while on the 31st we were back in Strathspey and used my mirror-pole to check a failed nest, which held one unhatched egg. In the afternoon we drove north to meet Colin Crooke and ring two late young at one of his nests near Inverness. The next morning, after more interesting discussions about Catalonia and the merits and techniques of translocating ospreys, I drove Manuel to the airport. He suggested I come to see potential locations in his area but, sadly, the organisation he worked for changed its focus and osprey conservation was downgraded. He was very disappointed, and so was I: another chance missed.

## Tuscany, Italy

In 2003 I received an invitation from Giampiero Sammuri, the then President of the Maremma Natural Park, to visit Tuscany to discuss the potential for reintroducing ospreys to Italy. On 29 March 2004 I was collected from the airport in Rome by Alessandro Troisi, whom I had met at the Urbino meeting, and he had stayed with us in Moray, to view our ospreys, with his friend Flavio Monti. Alessandro was (and still is) an osprey enthusiast, a superb bird artist and a member of the Italian osprey working group, and that evening we stayed at a friend's house. The next morning he drove me to La Maremma Natural Park in Tuscany, where I met the president and his staff. There I also met my old friend Jean-Claude Thibault from Corsica and his colleague Jean Marie Dominici from the Corsica Natural Park. I had visited Corsica in 1977 to view breeding ospreys, and later Jean-Claude invited me to a very interesting and enjoyable workshop on Mediterranean ospreys at Galeri in Corsica in late September 1991.

After a warm welcome at the park headquarters, Alessandro gave an excellent presentation on ospreys in Italy and his hopes for reintroducing the species. The last pair nested in Tuscany in 1929, the last elsewhere in mainland Italy in 1955 and the last in the islands in the 1960s. We then set off with the staff to be shown around the park. We first visited the River Ombrone, where the banks of the river looked really good for ospreys. It was a big, slow-moving river with some large, dead trees, with three artificial nests that they had built, complete with dummy ospreys. It was a good site for an osprey reintroduction and I was told that it was a protected area with no access. In the afternoon we had a workshop, where I gave a talk on reintroductions and we had very interesting discussions, which recommended that young ospreys would be collected in Corsica and translocated to Tuscany. Soon after, the Corsicans returned home and I was settled in a small guesthouse.

The next morning I was collected at 7.30 am and then had time for some quick discussions in the headquarters before we set off to explore as much of the area as possible, often to overview the coast. With the permission of the landowner we drove up over the Col d'Forno, which gave great views, and then went on to the village, where we had lunch with Andrea Sforzi, the museum director from Grossetti, who was carrying out PhD research on fallow and roe deer. Pietro Giovacacchini arrived and we set off on private tracks to the big marshes behind the sea, really excellent wader and wildfowl habitats and good for ospreys. Further south we visited another extensive marshland, which also contained two fish farms, growing sea bass and dorado. From the visitor centre we had further views of excellent osprey habitat, and I also saw a great spotted cuckoo, which was a nice bird for me. Early next morning, as I walked down the River Ombrone towards the sea, a male osprey flew to perch in a big poplar tree on the bank and, 10 minutes later, another was flying up the river, nearly a kilometre distant. I thought they were good omens for the project, for I had no doubt in my mind that it was good breeding habitat for a restored population. The fact that many migrant ospreys passed through gave the chance of extra female mates being available for translocated males. My assessment was that the Maremma Natural Park was an excellent location in which to restore breeding ospreys in Italy, using young birds translocated from Corsica.

I maintained contact with the osprey group and in December 2004 sent a detailed technical note to the Natural Park and to Corsica concerning all aspects of a translocation – collection, transfer, hacking cages, feeding of young, monitoring, release and post release. In early 2005 Jean-Claude wrote to me with an invitation to be in Corsica helping with the first collection of young, but arrangements were not moving that fast. On 7 August 2005 we met the Italian group at Rutland Water when we were preparing to release young ospreys and were able to show them many aspects of the work, including how to fit tiny VHF tail-mounted transmitters. Later, Tim Mackrill and Barrie Galpin went out to Italy to examine the chosen release area.

In February 2006 I was asked to send Andrea Sforzi a signed hard copy support letter, which Jean-Claude suggested was necessary to send along with the application to the authorities in Paris. Soon after, they received permissions and the first six

First osprey nest in Italy following the reintroduction at La Maremma. (*Photo by Alessandro Troisi*)

young were flown by helicopter from the Scandola Marine Reserve in Corsica. Tim and John Wright were invited to help with the release in 2006 and 2007. In total, 33 young were translocated between 2006 and 2013. In 2010 three pairs built nests, which was an excellent start; all three pairs involved translocated males and migrant females. In 2011 one of the pairs reared two young, the first in Tuscany for nearly 80 years. This new population continues to grow, with three pairs rearing nine young in 2017, and five pairs with eight young in 2019.

## Basque Country, Spain

In the summer of 2007 I satellite-tagged a breeding female osprey near my home in Moray and named her Logie, after the nearby primary school, so that they could follow her travels and learn about the life of an osprey. She was the first osprey that I had tracked with the new-style GPS transmitters, which gave incredible detail of migrations. She wintered in the Bijagos Archipelago in Guinea-Bissau and, after spending the whole winter on a small island, set off on her spring migration at midday on 12 March for a flight of 5,200 kilometres. She made good time through Africa and on into Spain, but when she reached the Basque Country she ran into really bad weather, westerly winds and heavy rain. She had to settle down and wait for a break in the weather, staying from 29 March to 7 April on the lovely Urdaibai Biosphere Reserve, an estuary just north of Bilbao on the Bay of Biscay coast. Her locations were so accurate that I decided to see if I could find someone to look for her, and an internet search gave me an email address for Aitor Galarza, a biologist working for the local environment department. Within a few hours Aitor had found and photographed

Logie eating a fish on the tidal flats. We kept in touch, with me sending him the coordinates of where she roosted each night in woods around the estuary and her fishing locations during the day. It was a contact that would lead on to another reintroduction.

That summer, on 18 July, Aitor arrived in Scotland to learn more about breeding ospreys. That evening I showed him the local nests, including Logie's, with her two young, a sharp contrast with how he had last seen her. The next day we went to Loch Garten, as well as to see another six pairs in that area, and on the 20th it was fieldwork again on my local Moray ospreys, followed by good discussions on the possibility of us helping with a translocation of Scottish ospreys to the Basque Country. The next day I showed Aitor the feeding areas of our ospreys at Findhorn Bay, the mouth of the River Findhorn, which is roughly the size of the Urdaibai estuary.

In early October I was invited by Aitor to carry out an assessment of Urdaibai Biosphere Reserve and the surrounding coastline for a potential reintroduction of ospreys, for they had not bred there for a very long time. We had a really exciting time exploring all the areas where Logie had stayed, even going by boat to the tall sea cliff where she settled to roost that first night. We checked out the roost sites in the woods and discussed potential nest sites both in the woods and on the estuary. I was very impressed by the new visitor centre, which would make a really good headquarters for such a project, and then found a good place for a hacking site along the eastern shore of the estuary, looking out over the marshes and mudflats. We needed to unlock a chain to drive down a private track to get to a sloping field, which could hold the hacking cages. On the evening of 6 October, Aitor and his colleagues organised a presentation in Bilbao, at which I gave a lecture on reintroductions as well as on the ospreys in Scotland and their migrations. About 60 people attended and we had a really good discussion, followed by further talk over dinner in the town. On my last day in the field we drove west to check out an extremely good area for ospreys called the Santona estuary and marshes. At two places where we could look into the water from a dock, I saw the biggest concentrations of mullet I had ever seen; no wonder young Scottish ospreys like these estuaries of northern Spain and France.

In autumn 2011 we were tracking three young Welsh ospreys to Africa as part of the BBC *Autumnwatch* programme. One of them pitched up on the hills of the Basque Country, so on 12 September we met Aitor at Bilbao airport, drove up into the hills and walked to the exact location in a wood. This was where I did an interview with Aitor about osprey migration, while later that day, and the next, Lucy Smith from BBC Bristol and her team filmed around the Urdaibai location. It gave me a chance to catch up with Aitor's news, and I heard that he and his team were still working on the proposal and had already put up a range of artificial nest platforms in the hills and in the estuary. The estuary nests – one in a pine and the other on a pole – were excellent and had been used by migrant ospreys, but none had stayed.

The following spring, Aitor emailed to say that they had funds and authorisation to begin the reintroduction of the osprey at the Urdaibai Biosphere Reserve in 2013. It would be financed by the Department of the Environment of the Biscay Regional Government and would be carried out by the Urdaibai Bird Centre within the Aranzadi Society of Sciences. Obviously, Aitor needed to know the willingness of the donor countries to provide osprey chicks for the project, and he asked if it might be possible to get chicks from Scotland during the period 2013–17. I immediately replied to say that I was in favour and that I would discuss his request with the authorities in Scotland. In the summer I was able to reply that, after our Andalusia Project meeting in the spring, it was decided that they did not need the extra nine birds from Scotland, which would allow us to find donor young for the Basque Country project. I wrote to Ben Ross, the licensing officer of Scottish Natural Heritage, cancelling the final part of the Andalusian licence and recommending that we assist the Basque Country, which I totally supported and regarded as excellent conservation cooperation between the two regions. I suggested to Aitor that he and the Biscay Regional Government put together a formal request to Scotland and SNH, which wanted clarification on the need for recapture of young after release to fit satellite transmitters. I submitted an application to collect the young ospreys in Scotland and translocate them to Spain, and the necessary licence from SNH to collect 12 young per year for five years was issued in February 2013.

The Urdaibai collection started in the usual manner on 5 July, revisiting nests which I had already been monitoring since April. I drove north that day to Caithness and from a regular nest on a farm I collected one of two young, and in the afternoon one of the middle young from an exceptional nest of four near Kinbrace. As I drove home I met Brian Etheridge, who'd also collected one chick for the project. The next day I was in Strathspey, but the chosen nest had only one young, so we went downriver to my east Moray study area and collected three from three pairs. In the evening, Aitor arrived and during the day I collected four young from my local area, and by midday on 8 July we had collected our full complement of 12 young. Jane Harley, our veterinarian, arrived to inspect them and authorise the health certificate, and I completed the Balai Directive document for travel authorisation.

The next morning Aitor and I were up at 3.45 am for the drive to Aberdeen airport, where the osprey chicks were fed with fish before being handed over to the transit people. Aitor and I caught the BA flight to Heathrow. The previous day I had phoned Tristan Bradley at the Animal Health Centre and he collected us from Terminal 5, while the birds were soon delivered to his facility. This was an extremely useful arrangement that we used over the next five years, and allowed us to feed the birds fresh trout before the next stage of their journey. We arrived at Madrid in the late afternoon and were collected by a small lorry from the Environment Department for the drive to Urdaibai. Unfortunately, the lorry broke down and we had to be rescued by two of Aitor's colleagues in small vans, who drove us through the night to deliver the birds safely to the hacking station. I finally got to bed at 8.30 am and then spent two days with the team at the hacking centre, helping to pass on our hacking techniques and getting an opportunity to have a closer look at the local habitats.

The first young were released on 27 July, and by 8 August eight of the young ospreys were flying free at the estuary. They often perched on the osprey nest out in the marshes, which was very close to the channel, but it was still a surprise to me to hear that one of them caught a grey mullet just five days after release. The last four were released on 10 August, and sadly one of them fell through the top of a large eucalyptus tree and damaged its leg. The first young departed on 31 August.

Over the next four years we completed the collection and translocation of 60 young ospreys to the Basque Country. In Scotland we were very grateful to the Forestry Commission and private landowners for allowing us onto their land to collect young ospreys. Jane Harley from Strathspey Veterinary Centre carried out all the inspections; I organised all the official

Newly released ospreys outside a hacking cage in the Basque Country.
(*Photo by Urdaibai Bird Center*)

paperwork and transportation; Animal Couriers organised the travel details for import to Spain of the birds; Gabriella Tamasi of BA Cargo was very helpful with flights, and Tristan Bradley and his colleagues at the Animal Health Centre at Heathrow airport offered their facilities free of charge for the transfer, and bought fresh rainbow trout for the birds. We always chose the second week of July for the collection because we knew the birds were at the prime age; they were kept in specially made compartments at my home and fed by us on rainbow trout, usually donated by the Rothiemurchus fishery. We sent out 12 birds in 2013, 11 in 2014, 13 in 2015, 12 in 2016 and the final 12 in 2017.

The BBC's *The One Show* asked if they could film the project in 2014, so on 7 July we were joined by Mike Dilger, the presenter. The next morning the film crew were with us as Ian

Perks climbed to a nest in an old Scots pine and lowered two young ospreys, which I measured, weighed and ringed. One was returned to the nest and the other collected for translocation. On 1 August I met Mike and the BBC crew at Brussels airport to fly to Bilbao. Aitor took us to the hacking site so that we could see the birds from Scotland, which were looking good and ready to be released. That afternoon the BBC filmed around the estuary, and at dawn next day we were back at the hack site, where the first cage door had been lowered. The cameraman was installed in a hide, while the rest of us were in and around the project caravan. After several hours the first young osprey flew free, making for a dramatic piece for Mike's film. It brought the reintroduction of ospreys to an evening audience of several million people.

In autumn 2015 I took Ian Perks and Fraser Cormack, our expert tree climbers, to Urdaibai as a thank you for collecting young ospreys. On the first day we went to the inland lakes and saw a 2013 translocated osprey: N3, a male. We saw two others at the next lake but they could have been migrants; we also discussed nest building potential. The next day we were at the hack site and there were still five juveniles remaining and one of the adults, while on the last day Aitor showed Ian and Fraser the estuary and the island offshore from a boat. In May 2017 it was the turn of Brian Etheridge, who had helped with collecting osprey chicks, to visit the release area. We were there for three days and saw the male osprey resident on the nest in the Urdaibai marsh but, sadly, no female had lingered. When we visited the inland lakes we found a white stork pair had taken over one of Aitor's osprey nests. On the last day we saw the 2013 male osprey in Santona marsh but no nest pole had yet been erected. The same was true at Santander, where, on 13 May, we were shown the osprey male, which had arrived from Urdaibai on 11 April, but the nest pole was still in the yard. We encouraged Carlos Sainz to get it erected and heard later that it was in place on 28 May and that just two days later the male had attracted a female. Nest building and mating occurred but no eggs were laid; the pair stayed to 21 September. The continual provision of artificial nests is very important to encourage breeding in areas where breeding ospreys have become extinct. Thirteen GPS 30 gm satellite tags were attached to young ospreys – 5 in both 2013 and 2014 and singles in 2015, 2016 and 2017. Five were females and eight were males; none of them survived to return to breed,

Mike Dilger and author with Scottish osprey bound for Urdaibai Reserve near Bilbao, for feature on BBC's *The One Show*.

Nest platform being built in Urdaibai marsh by staff and volunteers. (*Photo by Aitor Galarza*)

which supported my advice that translocated young should not be satellite tagged. With every project there's a balance to be struck between reintroduction conservation, public engagement and research; my view is that the first should be dominant.

In 2015 four two-year-old translocated males were observed in northern Spain, three at the Urdaibai estuary – where one took over the main nest and carried out nest building – while one was identified further west in Asturias. In 2016 four males were identified at Urdaibai, including the three males from 2013, again one busy at the main nest, where several times passing females visited briefly but did not stay. One was again seen in Asturias and a male was observed at the Courant d'Huchet reserve in France. In 2017 a pair were established at Santander Bay, and another pair, a male from 2013 and a female from Corsica, were at Courant d'Huchet. In 2018 the French coast pair reared two young but the Santander pair failed during incubation. In 2019 three males were seen at Urdaibai, the usual 2013 male at the main nest plus 2014 and 2015 males, but they did not succeed in attracting females to stay with them. The Santander pair were present all summer but apparently did not lay eggs, while the French coast pair had just one young. A new male from the 2017 cohort arrived in July and even attracted a migrant female in September. Some of the sightings of Basque Country releases have been seen very briefly at sites and then were not recorded during the rest of the summer, which gives hope that there might be undetected pairs in the greater region.

## Switzerland

In August 2011 Wendy Strahm and Denis Landenbergue wrote to me saying they were keen to restore breeding ospreys to Switzerland, as part of the centenary celebrations of Nos Oiseaux, the ornithological society in the French-speaking part of the country. By coincidence, the last recorded breeding pair in Switzerland was a hundred years before, in 1911. I replied positively and was invited to visit Switzerland from 10 to 15 February. An itinerary had been planned to visit lakes and wetland areas in western Switzerland, mainly in the cantons of Jura, Vaud and Fribourg, with some visits in the cantons of Neuchâtel, Bern and Geneva as well as in neighbouring France.

On 11 February we had a meeting with Oliver Biber, the president, and 13 other committee members of Nos Oiseaux at Yverdon, where I gave a presentation on osprey recovery. After a visit to a frozen Lake Neuchâtel, we drove to the French Jura and met Michel Juillard, Alain Rebetez, Rolf Wahl and Georges Contejean. The next morning we saw the reserve at Damphreux, which was a good location for a release site, and other potential areas. The next day, we were guided to many parts of the south shore of Lake Neuchâtel; the habitat was excellent, but natural nest sites were scarce. On 14 February we explored Lakes Morat and Bienne, both of which were excellent, with varied countryside, and there were potential breeding sites. Michel Beaud later took us to Lake Gruyère, where he had constructed an osprey nest in a conifer. The following day we visited the reserve at Gros Brasset, at the end of Lake Geneva.

A month later I sent a detailed assessment, which concluded that Nos Oiseaux's region of Switzerland – especially its fishing habitats and food availability – was suitable for breeding ospreys, although there was a general shortage of suitable nesting trees. Lakes, canals, dams and rivers were rich in numbers and species of fish, but natural nesting trees were scarce due to forest management. I saw many places where nest poles and artificial nests built in the tops of prominent trees could be used by ospreys. The hacking of young ospreys translocated from a donor population – Germany, for example – could restore a breeding population to Switzerland. In the areas I visited, I could easily visualise over 50 breeding pairs of ospreys.

A small group, including Wendy, Denis and Michel, put together a feasibility report on restoring ospreys with recommendations on best sites and the potential funding requirements. The committee of Nos Oiseaux was supportive but other groups, including Swiss BirdLife and the German-speaking national bodies, were opposed to the plan. I gave advice and support when needed, and was asked to return in January 2018 to help with several meetings to discuss their project. It was a pleasure and I found that Rolf Wahl, my osprey friend from the Orléans forest, would also be with us. On 18 January I again had the pleasure of meeting Luc Hoffman, the veteran ornithologist who has done so much for the research and conservation of wetlands and wetland birds throughout the world. In the evening we attended a meeting at the headquarters of IUCN, called to discuss the project. Most people were supportive but a small number of younger birdwatchers were very opposed, so it was an interesting evening.

The more important meeting was on Monday at the headquarters of Sempach, the national ornithological NGO, who were not in favour. There we had a useful discussion about ospreys with seven members of staff, including the director. The principal objection was that osprey did not feature in the list of 50 species of concern in Switzerland and that a reintroduction might detract from that work. The list had been chosen by Sempach and others, but we pointed out that lists are subjective despite scientific criteria, and the suggestion that any funding which went to an osprey project would deny support for another species was incorrect. A potential funder for an osprey project would not necessarily gift money to a less iconic species in need of conservation, but might instead sponsor the arts or sport, for example. In the end, opposition lessened, but I started to understand the local tensions between the cantons and between French- and German-speaking parts of Switzerland.

We again visited the locations most likely to be suitable for a release site, but Michel Beaud had come up with an even better option: in a prison at the junction of three large lakes – Neuchâtel, Morat and Bienne. Michel knew a person who worked at

Bellechasse State Penitentiary, and we were invited to inspect the large farm that belonged to the prison. It proved to be an ideal location, obviously private and secluded, but the open landscape, surrounded by farmland and strips of mature trees, and the close proximity of three very good lakes made it excellent as a hacking site. Martin Herbach, the farm manager, was interested; the prison authorities gave permission, and Martin also arranged for the prisoners to build the hacking cages in their workshops.

It had been another very interesting visit to Switzerland and the start of the project was in view for Wendy, Denis and Michel, although they still had a lot of work to do in finishing the plans and obtaining licences and donor young, as well as the necessary funding. They received authorisation from the Canton of Fribourg in April 2013, but authorisation from the Swiss Federal Office of the Environment did not come through until 28 May 2014, so the start was set for 2015.

In May 2015 we had a family long weekend with Wendy and Denis in the mountains, and one day we visited the hack site with Michel Beaud. Martin showed us the cages in the workshops and they were excellent; next, we looked at the chosen release site, which I thought was very good. I pointed out the best places for artificial T-perches and low feeding nests for the released young ospreys, and also some sites for a couple of nests built on tall poles. They had been offered accommodation for the project team a short distance across the fields. It was great to see such progress. A little while later they were having difficulties finalising licences for collecting young in Germany, so they asked for my help. I approached Scottish Natural Heritage and a licence was approved for one year to collect six donor young so that the project could proceed.

Despite still helping the Basque project, we were able to provide six young, thanks to Dave Anderson collecting three of them near Aberfoyle. On 13 June I drove with them to Aberdeen airport, having a slight worry when I heard on the car radio of possible bird flu in Lancashire, but soon the birds were at Heathrow. Tim Mackrill met me and we went round to the Animal Reception Centre, where we fed fresh trout to the young. The next morning I was up early for the flight to Geneva, where I was collected by Denis and then met Wendy at the cargo building. The ospreys had arrived and I was escorted through to customs, where the customs officer and a veterinarian examined them; they were very interested in the project and wanted to learn more. We drove to the hacking cages and settled them in three cages with fresh cut-up fish. Michel and Adrian Aebischer were there and we discussed the transmitter equipment. The next morning I went with Wendy to get fresh fish, caught that morning by a fisherman at Lake Morat. I could see they would be well fed.

Hacking cages for ospreys with feeders and perches in front, at the Swiss Reintroduction Site near Lake Neuchâtel.

Later in the morning Martin arrived with the prison governor and both were very impressed by the new arrivals. I left the Scottish young in very capable hands, monitored from a caravan 350 metres from the hacking cages using CCTV.

In early August I returned for two days to check the young and to help fit the tiny VHF transmitters to tail feathers, important for monitoring the released young in the first couple of weeks. Michel, Adrian and Daniel Schmidt from Germany were also there and we had very useful discussions with the project team. On the second day we surveyed the surrounding forests for suitable nesting trees, which Daniel would build over the winter. The first three young were released on 8 August and the other three on the 11th, and quickly they were eating fresh fish from the nest platforms, although kites also took advantage of the food. Four birds survived the month-long post-release stage and migrated between 2 and 9 September. Unfortunately one was shot and injured in Algeria, but it was looked after and released again on 11 October.

During the winter of 2015 the Osprey Project Team worked hard and finally got agreement to collect six birds in Norway and six in Germany; and this became their normal routine. In total, 24 young have been translocated from Germany and 25 from Norway between 2016 and 2019. In 2018 two ospreys from the 2016 cohort were identified in Switzerland, with one often present in the release area; he returned in 2019, when a 2017 young also returned. These birds have used the pole nests at Bellechasse, with occasional nest building and interactions with newly released young. One of the returning males was also identified in its winter quarters in northern Senegal. The project is close to success and just requires a female to join one of the summer resident males.

## France

In June 2000 I visited the Foret d'Orléans to see Rolf Wahl's study nests, stay with him for a few days and learn about the French ospreys. In 2001, after my field visit with Rolf, I was asked to write a discussion document in which I recommended an increase in artificial nest building outside the then range to encourage dispersal, and suggested that young could be translocated to several very good locations in other parts of France to expand the range and population size.

Twelve years later I attended a very interesting three-day International Osprey Symposium at the Orléans museum, where I met many old friends, and together we explored many interesting options. I gave two talks on ospreys, the second recommending a determined approach to the proactive management of ospreys in order to restore the southern European range. In France, the suggestion was to translocate young from Orléans to productive coastal and estuarine habitats with rich stocks of mullet. This talk led to Eladio Fernandez-Galiano of the Council for Europe asking me to prepare a document for them, which was published in 2017 by the Council for Europe. It was named the Plan for the Recovery and Conservation of Ospreys in Europe and the Mediterranean Region in particular.

At the Orléans symposium I spoke with Paul Lesclaux, who was very keen to try to restore breeding ospreys to the French coast south of Bordeaux. He and Florent Lagarde worked on their ideas and in January 2018 wrote to several osprey experts for advice, particularly asking for any evidence of impacts on donor populations. In February Tim Mackrill and I sent a detailed, positive response, along with our support, and as their project progressed we responded to queries on hacking cages, feeding and transport of young. Meanwhile, one of the Urdaibai males had, in the previous summer, relocated to their reserve and attracted a Corsican female, rearing two young in 2018. The first translocation of 10 young ospreys from the Orléans population was carried out in the summer of 2018, with the help of Rolf, and nine young were successfully reared and released, and all migrated. Another 10 young were translocated in 2019 and eight of them migrated. The breeding pair reared one young, and a new male from the Basque Country project, which Aitor's team reared and released in 2017, also settled in their area. The plans are to translocate another 30 young over three years. This project will also help the Urdaibai project to build up a new population of ospreys along the coasts of the Bay of Biscay.

Author with tame eagle owl during filming for a BBC TV programme on eagle owls, May 2005.

# 9

# ENDURANCE

## *Keeping hopes alive*

With wildlife conservation, more often than not, comes unfinished business: plans change, choices narrow, funds are not forthcoming and projects extend far into the distant future. The reintroduction projects included in this chapter encapsulate those difficulties and challenges. The species are varied: cranes, golden eagles, great bustards and eagle owls. Some projects are contentious and all are difficult.

They are not the only ones we've looked into. In September 2006 I was asked by Mark Avery of the RSPB to give a talk to their wardens' conference in Cheltenham about my ideas on reintroducing species. It was a good get-together and my talk stimulated a lot of questions. I was encouraged that the RSPB was taking a greater interest in translocations and reintroductions. In addition to those already underway, I outlined a range of species that I found interesting. Black storks, for example: populations in mainland Europe had increased in numbers and range in recent decades, and occasional vagrants were now regularly seen in Britain. We could wait for them to come – for they once did nest here – or we could be proactive and bring them over. I favoured the latter and thought that there were many woodland streams in the UK which would be suitable for the species. Once the beaver becomes more common, the black stork will be able to enjoy its most favoured habitat – beaver ponds.

Dalmatian pelican is another strong contender for action, for it is unlikely to get back here without assistance, although a single bird turned up a few years ago. Hazel hen would be another wonderful species to restore to the British Isles and it cannot return without assistance. Equally important for the future of nature in a damaged landscape is the need to enhance the viability of remnant and isolated populations of scarce creatures of all forms Too many are now trapped in nature reserves surrounded by agriculture and suffer loss of genetic vigour. The choughs of Scotland, for example, are restricted, with blindness of chicks becoming a problem, so it's essential that chough populations be mixed up by translocations. And, when we look, there are many others. It's time for bold and proactive action for nature.

## Common crane

I saw my first common crane, nowadays officially called the Eurasian crane, at Dunrossness in Shetland on 10 August 1972. They were scarce birds in those times and I was lucky to see it fly over Robin's Brae towards Loch Spiggie. My diary records its plumage and distinctive shape in flight before it dropped out of view. Cranes are big tall birds, larger than grey herons, and a striking sight. My next encounter with the species in Scotland was of a pair which had escaped from the Highland Wildlife Park in July 1982 and spent the autumn and winter living together in Insh marshes. The Park wanted them back so, on 5

November, a small party of us from the Highland Ringing Group arrived at Gordonhall Farm at dawn to meet Doug Weir. We set out our cannon net baited with grain on the cranes' favourite field and did not have long to wait before we heard their beautiful, trumpeting calls as they flew in. At the last moment, though, they turned away. Before long they were back, but would not go near the hidden net, although we even used the farmer's tractor to try to shepherd them. When they flew off, we moved the net, but they knew something was up and did not return until the afternoon, and again avoided our net. They may have escaped from captivity but they were certainly not tame. In the spring they became noisier with their calling and then left on 15 April 1983. Surprisingly they were not seen again, but two were back in the Insh marshes from 3 to 9 May 1987. Were they the same ones? We do not know. In mid-May 1988 I went to Teindland moors near Elgin in Moray to check up on a report of a pair of cranes in that district, which may have been there in previous summers. I found nothing, but it's a remote, quiet corner with few visitors and it would have been nice if the escaped pair had tried to breed in Scotland.

I got to know cranes well when I visited Sweden in the summer of 1983, for they were a regular presence in the marshes and forest bogs that I visited. I saw them again in following years, in summer in the north and in winter in Spain. From reading the history books I knew that they had once been a regular breeding bird in the British Isles but had been exterminated centuries ago because they were good to eat. The crane was well known in ancient times, for there are over 300 place names throughout Britain which include some reference to the crane, and in Scotland we have place names such as Cranloch, Cranstoun and Cranbog Moss. It's the commonest place name referring to a bird, in fact. As well as place-name and archaeological evidence, there are also many documentary sources and manuscripts that make it clear that the crane was a well-known breeding bird in Scotland, not just a winter visitor.

The crane evidently bred in Scotland in 1529 and earlier. The Earl of Atholl provided a banquet of 'cran, swan … and capercaillie' for James V when he went hunting in Perthshire. Cranes were also purchased for his larder as his 'Household Accounts' suggest that cranes breeding in Scotland were treated ruthlessly. In 1578 the Bishop of Ross wrote that cranes inhabited Ross and Inverness. The crane is also used in Gaelic proverbs such as 'Cha chluinn e glaodhaich nan corr' ('He cannot even hear the crane's creaking cry'), 'Ghoideadh e an t-ugh bho 'n chorr 's a' chorr fhein 'na dheireadh' ('He'd steal a crane's egg from the crane, and the crane herself at his heels'), and 'Fhuair mi nead na corra-dubh ann an cuil no mona' (an old verse of Port-a-beul, or mouth music, translated as 'I found a crane's nest in a hollow in the moor').

The crane was clearly a bird of great significance to our ancestors in Scotland; its trumpeting flocks returning in the spring, in V formations, would have been one of the great events of the yearly calendar for country people. To see and hear cranes is to fall under their spell, which is why I wanted to see this dramatic bird back in our country. In the mid-1990s I started to gather information, with the idea that we might try to reintroduce them.

In April 1996 I went to Minnesota to learn about osprey reintroductions for our project at Rutland Water, and beforehand I wrote to George Archibald, the director and creator of the International Crane Foundation, based at Baraboo in Wisconsin. I asked if I might visit after my osprey meetings to learn more about crane reintroductions, for I was considering one for Scotland. I received an enthusiastic response, explaining that such a project might be helpful to them in developing techniques for the endangered Siberian cranes in Russia. He suggested getting late-term incubated eggs from Sweden and transporting them to Scotland in an incubator, and then, after hatching, rearing the young in the habitats where they might breed in the future, with guardians wearing crane-costumes. He recommended fitting radio transmitters and teaching the crane chicks to walk behind an ultra-light aircraft and then to fly behind it, so that the cranes could be led on a flight from Scotland to Spain via France. In Spain they would join other flocks of cranes from mainland Europe and hopefully, when the time came, they would return to Scotland in the spring. Well, this was the sort of response that makes the heart race.

I drove from the airport to Baraboo on 30 April and gained 24 hours of concentrated crane knowledge. I visited crane city, where the first track we walked up was called Sibe road after the Siberian cranes, and through a big series of extensive and well-maintained pens, where I saw pairs of most species of cranes in the world, many of them breeding and viewable through screens. George was away chasing cranes in Asia, but his colleagues – Claire, Jeff and Kurt – answered all my queries about cranes and more, and sent me away with a remarkable set of scientific papers and reports. It certainly triggered in me a desire to see the crane returned as a breeding bird in Scotland. I left a series of habitat slides for George to view on his return, and when he sent them back he said there were plenty of habitats for cranes in Scotland and gave me the names of crane experts in Germany and Sweden.

On 18 September that year I made a trip to Hickling Broad to meet John Buxton, who owned this special area of marshland in

Norfolk. I had known John from the early 1960s when I helped him make a film about ospreys in Strathspey. It was a very good day, first of all talking with him and his wife in their home before John took me out onto the marshes, where I saw five of the seven cranes there. I learned from John that the first three arrived as sub-adults in 1979 and stayed through the winter. In spring they flew off north, but about two weeks later they returned, with much calling. In 1981 the first breeding occurred, but the single young was lost to a predator. The following spring they did much display and built nests in the big reed beds, walking out to feed in the rough meadows. There were very few, if any, tracks in the reeds, so it was difficult for the chicks to walk out, but a single young was reared and successfully flew. Colonisation was slow and predation of the young, believed to be mostly by foxes, was a real problem, but numbers started to increase by the end of the century.

After my visit I wrote to George Archibald with an update, explaining that our osprey project had taken longer to get licences and agreements than we had hoped and that crane planning had taken a back seat. I reported that I had met John Buxton in Norfolk, who had been disappointed to see how little encouragement he had received from the nature conservation bodies to increase the viability of the population. I informed him that I had broached crane reintroduction with various people in Scotland so that it could slowly mature in their minds, but had not received much enthusiasm, so 1998 was a more realistic start time. By that year, though, it was clear that I would find it very difficult to get permission to carry out a crane reintroduction.

In 2000 the osprey reintroduction to Rutland Water was going well and the first young was hatched and fledged the following year. I decided to resurrect the crane proposal and asked Zoe Taylor to help me with searching the records. We wrote a feasibility document, 'Proposal to Aid the Recovery of Eurasian Crane (*Grus grus*) as a Breeding Species in Scotland'. Our research covered the ecology and breeding habitat, which recognised that it was a recovering species throughout much of Europe after earlier centuries of persecution for food and habitat loss. The history in Scotland was well documented and showed it was once a common and well-known breeding species, to judge by place names, literature and records. In fact, it was once one of the most notable birds to ancient people. Since the Middle Ages there has been a marked decrease in the range of the crane all over northern Europe, with extinction in several countries. Cranes were common breeding birds in Britain until the 1600s and in Ireland until the 1300s, with large migratory populations in Finland, Sweden and Norway. The crane was a rare migrant to Scotland in the first half of the 20th century, with sightings of cranes becoming more frequent from 1969 onwards. We analysed the data over five-year periods per month, noting the numbers arriving in different seasons and the variations in the length of stay – some were long-stayers of over 80 days. We also analysed the records for different counties, with Aberdeenshire, Caithness, Orkney and Shetland being favoured.

In our feasibility report we outlined the build-up of a small natural breeding population of cranes in Norfolk and explained that our investigations showed that cranes would be able to breed successfully in many areas of Scotland. The sticking point, though, was how to do it: the collection, incubating and hatching of young cranes was reasonably straightforward, but the rearing of the young cranes by carers dressed in crane uniforms and subsequent imprinting in flight on a micro-light aircraft was daunting and expensive. We wondered if there was another way.

In April 2001 I visited Brandenburg in northeast Germany to attend an osprey conference, and while there I decided to find out more about cranes. We saw many pairs settling down as we travelled in the countryside and it was interesting to see them in small, marshy areas and wetlands within a farmed landscape, where people accepted them as very much part of the scenery. The crane's arrival from Spain is a real sign of spring returning to northern lands. I was fortunate that, on 6 April, I was able to visit Wolfgang Mewes of the European Crane Working Group and had some very good discussions at the National Park headquarters at Lake Müritz. We talked about the ideas that I had for restoring cranes to Scotland, including whether these cranes should be migratory or if they would become semi-resident because of the milder winters in the British Isles. The North German cranes traditionally migrated to Spain, but Wolfgang told me that in recent years, with warmer winters, some birds had even stayed behind in Germany, while others had migrated only as far as France. He held a similar view to me: that in the extreme west of Europe and in the British Isles there would be no need for cranes to migrate in the future and, in fact, they may not have done so in the centuries when they were breeding here. They may have just gone to coastal areas and across the water to Ireland. We discussed techniques for reintroduction, as well as the special problems of catching sufficient numbers of young and how to acclimatise those young to adult birds for learning purposes.

I had Mikael Hake with me that day. Mikael is a Swedish osprey friend of mine, but at the time he was employed researching cranes in Sweden. We wondered why one could not rear a group of young cranes in Scotland and release them, and why they definitely needed to be with adults if they were not going to migrate. To us, that seemed a way forward, but Wolfgang thought that it might not work. Mikael's view was it was worth trying because it would be a much easier process. Two years later I attended the fifth European crane conference in Sweden, from 10 to 13 April 2003, where I again met Mikael Hake and Oliver Krone at the Flämslätt conference centre in Västra Götaland. The conference was organised by the Swedish Crane Working Group and the wildlife damage centre at Grimsö Wildlife Research Station, where Mikael worked. There was a brilliant series of lectures and really good discussions with many crane researchers from all over Europe and beyond. On 12 April we had a field trip to the famous crane observatory at Lake

The famous spring crane gathering at Lake Hornborga in Sweden, April 2003.

Hornborga, where a memorial to the founder, Per Olof Swanberg, was unveiled. There were thousands of cranes there that day, for it's where the birds congregate in early spring before heading off to nest in many parts of Sweden. It is now a spectacle enjoyed by thousands of visitors. I had conversations that day with friends about my hopes and, although the transport of eggs was feasible, my idea of catching half-grown young was considered more demanding. Oliver told me that he and friends helped with ringing young cranes and they were amazingly difficult to catch in boggy land, ponds and lakes. It would be a real challenge to catch 20 in a few days. Back home, I pondered what I had learned and, sadly, put the idea aside for the time being.

Ten years on, in late 2013, Justin Prigmore of the Cairngorms National Park asked me to have another look at the idea, and in 2014 I produced a report on 'The Feasibility and Desirability of Crane Restoration in the Cairngorms National Park'. My main conclusions were that there was suitable habitat and food for breeding cranes in the Cairngorms National Park and sufficient areas to host a viable population. I also considered that a restored population would benefit the national status of the species and, additionally, would introduce a very special, totemic bird to the avifauna of the park. I also thought that cranes breeding in the Cairngorms would not need to migrate long distances but could winter within Scotland. I considered that the species would have a positive impact on the nature of the park and on human enjoyment and ecotourism, and saw no negative impacts. I thought there were several options: one was a full-scale active recovery, like the Great Crane Project in England, which involved importing eggs and hatching and rearing young, while the other was the encouragement of wandering cranes to settle in suitable habitat by using decoy captive cranes. I also thought it was necessary to have a better feel for the priority of restoring cranes, and that my document should be used to bring together potential partners and to work out costs and a way ahead.

By 2014, the small crane population in East Anglia had increased and spread to Suffolk and Yorkshire, and the UK Crane Working Group reported that there were 18 pairs breeding which reared a total of 12 young in 2014. There were also two probable pairs, two possible pairs and another seven non-breeders. Additionally the Great Crane Project, a joint venture by the Wildfowl and Wetlands Trust, RSPB and the Pensthorpe Conservation Trust, started an ambitious recovery project in Gloucestershire and Somerset. They translocated eggs from Germany and between 2010 and 2014 released 94 young cranes. The first pair nested in 2013, and the following year there were three pairs on the Somerset Levels and two pairs at Slimbridge. In Scotland cranes had started to nest in northeast Aberdeenshire, with one pair in 2012 rearing a single young, rising to two pairs nesting in 2014.

On 6 January 2008 a flock of 11 cranes were seen to fly in off the sea at Lossiemouth in Moray, and on 11 January they were found feeding in farm fields at Pitgaveny, near Elgin. I went along immediately that morning and enjoyed watching the cranes feeding across the stubble fields; there was a pair with one young within the group of 11, which were feeding by turning over old straw to get at barley grains. They all stayed until at least 19 February, while the family party remained until 11 March. I remember thinking whether it would be worth trying to encourage them to stay by using some decoy cranes – for I still thought that technique might work with this sociable bird – but where would I get decoy cranes quickly?

In late September 2013 the national park had another try at stimulating the idea of reintroduction. We visited Glenfeshie with Thomas MacDonell, and also thought about using the

Highland Wildlife Park grounds. On 21 January Justin Prigmore and David Hetherington of the Cairngorms National Park took me with them to meet Hywel Maggs of the RSPB in Aberdeenshire. He very kindly showed us the crane breeding area, where there were now three potential breeding pairs. Hywel told us that breeding success was not as good as had been hoped, due mainly to fox and possibly badger predation. I thought predation of eggs and young might be a real problem in the national park. It was great to see the small Scottish population breeding and slowly increasing, and it would be good to consider ways of increasing and spreading the population, but I decided that I had tried over many years, had not succeeded and would give it a rest.

## Golden eagles in England

In February 1984 I was lecturing to members' groups of the RSPB in the north of England and, on the 18th, my colleagues took me to see the golden eagles in the Lake District. We walked up to the eagle viewpoint above Haweswater where I was shown the different nest sites in Riggindale, but there were no birds present, apart from a couple of carrion crows. On the way back along the lakeside we had good views of the pair of golden eagles flying further down the lake, being escorted by four ravens before a peregrine joined in the chase. The landscape was dreadfully overgrazed by sheep and – to me – extremely impoverished for wild prey for breeding eagles. A golden eagle pair was found in 1957 at Haweswater, but the first eggs were not laid until 1969; in 1970 an eaglet was fledged, the first in England for two centuries. Between 1970 and 1996 a total of 16 young were fledged, while a second pair bred from 1975 to 1983 and reared four young. The female of the last pair died in 2003 and the male was there alone until 2016.

My friend Brian Little in Newcastle often told me about the golden eagles that he monitored in Northumberland and their precarious position there. His notes record that a pioneering pair was killed in 1950 and that the next was a single in 1972; nesting occurred from 1975 to at least 1996, and 15 young were reared in that time. The last resident was in 2007, when the old female, which had lost its mate years before, moved over the Scottish border to the nest near Peebles, where the female had been killed. I always thought that golden eagles should be more widespread in England, not just in the hills or moorlands, for in other countries some of the most productive pairs lived on low ground, where prey was more varied and plentiful. I had talked with Brian and wondered whether it would help if golden eagles were taken from northern Scotland and released in Kielder Forest. He said that he would arrange a meeting to discuss it. I drove down to Northumberland on the evening of 12 October 1995 and stayed with forester Bill Birleton, in Bellingham. The next day we joined Brian Little and Ian Newton at the Forestry Commission office for Kielder Forest and had a good meeting with the Regional Forest Manager, Graham Gill, whom I had known in Scotland. I then wrote a proposal to boost the Northumbrian golden eagles, using translocations of 20 to 60 young eagles from Scotland, but we were unable to move it forward.

Many years later, in January 2009, a meeting of eagle experts convened at the Coignafearn estate, near Inverness, to discuss a range of innovative techniques to extend the range of golden eagles in Scotland and England. They included actions such as saving eaglets from broods of two in areas of poor food availability and fostering them in the best nests in rich food areas for subsequent translocations, and moving full-grown birds. After talking with Brian Little at a meeting of the Highland Raptor Study Group, I wrote to SNH requesting advice on taking these initiatives forward, but these were contentious ideas and I received no reply. In 2015 I wrote to the journal *British Birds* about the failure to restore England's eagles, saying that I found it truly worrying for the future of nature that although there are people capable of reintroducing eagles to these places now, such a project would attract opposition as well as demands for a massive science base, years of feasibility studies and stakeholder meetings, and then – possibly – a small trial. We just had to be bolder.

In 2017 we then recommended testing the effectiveness of translocating sub-adults by live-trapping, translocating and satellite tagging up to 12 golden eagles. By this time we were recommending translocation to places as far south as Hampshire and Devon. The Scottish Borders golden eagle project started at this time and my Foundation was later heavily involved in the proposal to translocate Scottish white-tailed eagle to the Isle of Wight. The worthy aim of restoring golden eagles to a range of locations in England is still on the wish-list, but with luck the work we have already done will help this exciting plan – to restore England's golden eagles – to become reality before long.

## Great bustard

In 1970, when I was warden at Fair Isle Bird Observatory, Gordon Barnes of Setter Croft reported that a great bustard had arrived on Fair Isle on 11 January. It was first seen in fields in the south of the island and then moved further north, where it was caught at night five days later, by Gordon, because it was very emaciated. He fed it on cabbages and the occasional dead mouse in a shed at his croft, where I saw it on my return to Fair Isle on 19 January. It gained weight in captivity, was ringed and released on 24 February, but unfortunately it did not thrive in the wild, so was recaptured at night on 5 March. I had heard of a project to restore the species to Salisbury Plain, and George Waterston put me in touch with the organiser, the Hon. Aylmer Tryon, in London. Thus started a plan to send it south, while Gordon continued to feed it cabbages in captivity. On 6 April I sent it out to Sumburgh in a big box, from where British European Airways

Great bustard at Fair Isle with crofter Gordon Barnes in 1970. After careful recuperation the bird was sent to Salisbury Plain to join a reintroduction project. (*Photo by Dennis Coutts*)

flew it south to London. It was released into the recently created bustard breeding enclosure at MoD Porton Down in Hampshire.

Five years later I was at the Natural History Museum in London to do some filming for a television programme, and I had time afterwards to have a long chat with Aylmer Tryon about his great bustard project. He told me that the Fair Isle bird was still alive but had not bred. Originally we had thought it was a young male, but it proved to be a female and lived for 11 years.

In 1980 I met a group of East German birdwatchers in Mongolia, and found that one of them, Max Dornbusch, was in charge of their great bustard project. He told me that when farmers disturbed a female from a nest on farmland, they were encouraged to collect the two eggs for incubation at the great bustard centre. These birds were released into the wild and, in his view, the one that turned up at Fair Isle was almost certainly from their project. In late May 1984 I visited one of the East German group, Werner Eichstädt, and had a very interesting four-day visit. One day, between Pasewalk and Löchnitz, we looked for great bustards in the farmland but without success. Werner confirmed that farmers were encouraged to collect eggs from nests when they had flushed a female, which often resulted in desertion; the reward was about £8 per egg for incubation, which resulted in the release of about 30 full-grown bustards each summer.

On 19 March 2006 I finally saw the great bustards on Salisbury Plain, when I met David Waters, who had resurrected the project in 1998 and was running the successful reintroduction. My visit was enhanced because his father, Estlin, was also there, and we were friends from long ago, having gone on a grey seal expedition to North Rona in the early 1960s. We went to the hide overlooking the big release area and watched four males and two females; I was told that there were also two males and a female outside the fence. It was a great chance for a really good talk about the project and I'm delighted that now the birds are doing well, following huge efforts by David and the Great Bustard Trust. In 2020, there were about 100 in the wild, with superb spring lekking and at least 21 females laid eggs. It is satisfying to think that our Fair Isle bird – fed on mice and cabbages in a crofter's shed – was one of the forerunners of this amazing project.

## The eagle owl dilemma

On 26 April 1984 I was in my RSPB office near Inverness when I received a telephone call from the liquidator for a quarry company near Elgin. Eagle owls were nesting in the unused quarry machinery, so Colin Crooke and I arranged to meet the manager, Alistair Reid, the next morning. He showed us the nest, on the chute of a stone crusher; there was eggshell, pellets and bones in a rough scrape in the gravel. Nearby we found two eagle owls that flew off and landed on the face of the Gedloch quarry. Eagle owls are very large, streaked brown, with ear tufts and orange eyes; they weigh up to four kilos and have a two-metre wingspan. They are very impressive birds. I saw my first in the wild in 1972, in the limestone hills of southern France, and another, 10 years later, in a rocky gorge in Mongolia, with occasional sightings in mainland Europe. One of my Finnish friends said that in his country they were called the 'assassins of the night'. It's a very apt name for this incredible owl.

At home in Scotland, I had often seen eagle owls in captivity; they were easily tamed and were much admired by falconers and the public. As a result I thought of them as tame birds, and guessed that the two owls in the quarry had escaped or even been released. I was sure that they were not wild birds. On 17

February 1985 I found that they had moved 2 kilometres to the Netherglen rock quarry, where I found a nest scrape at the base of a small sheer face on a rocky slope. When I returned on the 25th with Colin, we found the male perched in a holly tree on the sloping cliff, holding a plucked pigeon. When we checked the nest area we found many prey remains, including woodcock and red grouse. We waited until dusk and heard quiet duetting calls between the pair. On 7 March, a superb spring day, I found the female sitting well down, incubating eggs, in a nest under a juniper bush; on 14 April I went to the nest and the female was aggressive, vigorously bill clicking, and I saw a newly hatched chick and three eggs before quickly leaving. On 23 May, when the chick was well feathered, Colin and I went back and I ringed the chick at the nest while Colin fended off the female. She looked fearsome with her wings fully spread, swaying from side to side and clacking her bill just four metres away. The young one fledged in June but the other three eggs did not hatch. Sadly, one of the adults was found dead, by Hector Ogg, on the nearby road on 10 September, killed by a vehicle. Hector gave it to me, even though it was badly damaged, and I was able to identify it as the male. The female continued to live in the quarry for the next 10 years, sometimes laying eggs that failed to hatch and sometimes not; the last report of her, calling in the quarry at dusk, was on 19 December 1995.

In the 1980s, my work with the RSPB often involved incidents with birds of prey robberies from the wild; young peregrines were the most sought after, because of their financial value. During these investigations we might visit falconers' premises with the police, and it was not that unusual to find eagle owls, for at that time it was legal to keep and breed them. We heard of cases where eagle owls had escaped, and some were seen living in wild places. Some were clearly escapees, like the one

Eagle owl chick at the quarry nest in Moray,
May 1985.

seen at night on the roof of a fast-food outlet in Inverness centre, which suddenly swooped down to grab a herring gull that was eating chips on the pavement. The owl plucked and ate its prey on the roof, the feathers floating down into the street. At that time, nothing suggested to me that any of the owls might have arrived naturally from the continent.

Years later, in 2003, I was asked by Fergus Beeley, a BBC TV producer at the Natural History Unit in Bristol, if I would help him by presenting a Natural World documentary on eagle owls. He had done a lot of reading round the subject, and wanted us to explore whether or not eagle owls should be in Britain. Were they here long ago, and should they be returned? Although there were sightings in previous centuries, the official view was that the eagle owl was not a British bird. Fergus wanted me to cast my eye over the pros and cons by visiting breeding owls and people in mainland Europe and at home.

Our quest took us first to Sweden. On 27 August 2005 we drove from Lund in southern Sweden to Höör, where we met Staffan Akeby and went to visit the local wildlife park. There, the eagle owl recovery group Berguv was releasing owls. Since 1980, over 2,000 captive-bred eagle owls had been released in Sweden to restore the population, which had been decimated by hunters and farmers in earlier times. On the following day, while interviewing Staffan, I was honoured to release one of them, which had been bred further north. The next day, on our way back to the airport, we filmed with David Erterius at a rock quarry near Lund, where I could see one owl on the shattered rock face. David told me that the owls were used to the dynamite and quarrying activities; in fact, both men told me that eagle owls just fitted in with villages and towns, catching rats and crows in rubbish dumps, as well as rabbits, hares and hedgehogs. Of course, seeing an eagle owl in a town or village does not mean that it's an escaped captive bird: the first myth I'd wanted to test.

A month later we were in the Netherlands, visiting extensive quarries near Maastricht with local raptor expert Paul Voskamp. He took us to a deserted quarry which had been taken over by trees, creepers and bushes, where he showed me the eagle owl nest site in one of the cliffs. The next day he led us through an active quarry to a huge cement works, with a series of caves through the limestone, and we checked the nest sites. Of course, we saw no birds. He told me that it was also a very active bat roost. We did interviews at both places and, when scrambling up one of the wooded slopes, came across a predated adult common buzzard. Paul was not surprised, because the owls can kill a range of raptors; he had found 10 buzzards and 10 barn owls over three years as prey items. Buzzard numbers, though, had increased dramatically after hundreds of years of persecution, and near where he lived there might be a buzzard nest every hundred metres, so we both agreed that a predator-eating predator was just a normal part of nature. Paul told me that the first pair of eagle owls bred in 1997 and had colonised from Belgium or Germany. I also learned that the British view – that wild eagle owls live in wild and remote places – was simply not correct. In mainland Europe they were often close to people; they just need cliffs with ledges and enough to eat, and they can breed just about anywhere.

Our next exploration was in Switzerland, where we met Adrian Aebischer, a scientist studying eagle owls. It was a beautiful, cold, frosty day in mid-December, so we quickly left the village, with a packed lunch, and headed into a high valley where one of his radio-tracked owls was roosting. He located it with his tracker and we did an interview in the winter sunshine, with a few chamois on the cliffs behind us. We went on and reached 2,000 metres in the snow so that we could get great views across the Alps, for Adrian told us that some of the young eagle owls had dispersed over 300 kilometres and crossed 3,000-metre-high mountain passes into France and Italy. Some Scandinavian young have flown twice that distance. The next morning he took us to a big cliff among vineyards in the valley where a pair of eagle owls regularly nested on the rock face. It had been a big task for an experienced climber to get to the two young for ringing. While we filmed the interview, two choughs flew over our heads and crested tits called from the bushes. We had learned that eagle owls could fly long distances over a very difficult terrain. We also learned that Europe's eagle owls had bounced back from very low numbers.

In 2005 we concentrated on the British angle, for a pair of

eagle owls had been found breeding in Yorkshire. On the evening of 21 January I drove over a long and winding road in the Yorkshire moorlands, with a wonderful display of northern lights in the clear skies, to meet Tony Crease and his wife, and the BBC crew, near Richmond. A farmer had found the owls on Ministry of Defence land and Tony, a retired major and the MoD conservation officer, had confirmed them as eagle owls on 11 March 1997. We arranged to meet Tony before dawn and he walked us down to the edge of a valley; it was a cold, bright morning and we could immediately hear one of the owls hoot, an eerily deep double hoot – *oo-hu* – immediately answered by the mate. The male then landed in the top of a hawthorn bush by a small cliff on the other side of the valley. The nest was at the base of the cliff, and this was the only known pair breeding in Britain. Tony told me that 20 young had flown from there over the years. As the sun rose behind us, the owls fell silent. Later, with a telescope, I had really great views of the male owl perched in a yew tree, hunched up and roosting. We returned in the late afternoon: at dusk, the male flew back and forth past the hidden nest, and the pair were later calling to each other. It was magical. We returned in mid-May and from the slope opposite we could see the three owlets in the nest, so I accompanied Tony when he went to ring the young. While we were at the nest the female watched us from a tree but kept away. I could see that rabbits were the main food; in fact, it was all rabbits, Tony told me, for they were plentiful in the valley. Life was easy for these owls. I was allowed to ring one of the young while we did interviews. It was a real privilege to watch this nesting pair in their secret valley, with sheep and lambs grazing near them, while curlews and red grouse called in the background. We later learned that one of Tony's ringed young from 2004 was found electrocuted under a transmission pole, 200 kilometres south on a Shropshire farm, proving that the British ones can travel long distances, too.

We filmed interviews with many other people that year: people who had seen eagle owls near their homes, as well as owl keepers and ornithologists with very different takes on this bird, a controversial one in the UK but not so in mainland Europe. Members of the ornithological establishment were clearly very antagonistic towards eagle owls. Early published reports of the ornithological literature were scoured, and the 90 or so records since 1684 were rejected by the British Ornithologists Union's records committee. They argued that the species had not occurred in the wild state in Britain and Ireland, so the eagle owl was removed from the British list. Conservationists argued that

Tony Crease (Ministry of Defence Conservation Officer) with young eagle owl at nest site in North Yorkshire, May 2005.

eagle owls were a bad idea, for they would kill rare and protected species like hen harriers and corncrakes.

The film was broadcast on the BBC in mid-November and provoked a great deal of interest. Major debates were sparked on social media platforms, with widely varying views. I received a surprising number of letters saying how much people had enjoyed the programme and how much they had learned. A good few were from senior ornithologists who were persuaded by the arguments in the film. Like me, some did not agree with the present-day birdwatchers who expect pre-20th-century records to be documented as they are today. The early naturalists did not have cameras and superb optical gear, but they did collect specimens and knew what they were talking about. For others, predators eating predators was a natural part of the ecosystem rather than a threat. There was agreement that nearly all eagle owls in recent decades had escaped from captivity or been released, and that this should be tightened up, but many agreed that it was not impossible for owls from the greatly restored population on mainland Europe to disperse to Britain. Previous evidence for captive origin – such as living near people, being unafraid of humans, flying long distances and nesting in unusual places – was no longer seen to be sound.

Some people asked me if we should reintroduce the species, but I hesitated over that, for it's clearly controversial. I do object to them being regarded as aliens, for it is not possible to prove that every eagle owl seen in the wild is from captivity. Interestingly, an eagle owl found in Norfolk in November 2006 was investigated by analysing the stable isotopes in its feathers. The report, by the scientist Andrew Kelly and colleagues, concluded that an origin in Scandinavia, northern continental Europe or mid-continental Russia was consistent with their findings, but they could not rule out the possibility that the bird was reared in northern Britain, either in the wild or in captivity. It will be worth checking up on any eagle owl appearing at autumn migration times. Some day an individual ringed at a nest in mainland Europe may be found here, and then the bird will be back on the British list, where it belongs. Will that be a good thing? I think so, yes, for they are significant in the hierarchy of nature. The eagle owl can kill predators like buzzards, peregrines, ospreys and red kites, and also mammal predators like badgers, feral cats and martens, which often have few controls when numerous. Maybe it is time to get a bit of balance back?

White-tailed sea eagle. (*Photo by Mike Crutch/A9Birds*)

10

# OPTIMISM

## *Reintroducing sea eagles to Suffolk and Europe*

Now we come to some proposals that got away. There's always another time and place, of course, and it's salutary to remember that you have to have optimism to maintain a vision. The international sea eagle conference in Sweden in 2000 recognised that, mainly as a result of human persecution, the species was missing from the southern half of its European range. This was once a pan-European breeding species, south to the North African coasts, and the species is slow to recolonise lost range naturally. Reintroduction techniques for this species have been perfected in Scotland and could be used in other localities to restore the species' ancient haunts, which would increase the long-term viability of the species. The sea eagle, or white-tailed eagle, has a special and fascinating relationship with humans, stretching all the way back to early man and the Neanderthals. This chapter explores a range of proposals and ideas for carrying out its recovery, but none of them has yet taken place, although in the meantime one location has been recolonised. There is no doubt in my mind at all that this is unfinished business.

## The first attempt in England, Suffolk 2005

On 17 January 2005 I was at the annual winter osprey project round-up meeting at Rutland Water with Tim Appleton, Barrie Galpin and Tim Mackrill. After a very good meeting and a field trip to see four smew on the reservoir, Tim Appleton and I had a telephone discussion with Andy Brown, ecologist with Anglian Water, to discuss an idea of mine. Following the success of the osprey project, I felt that we should explore the idea of reintroducing white-tailed eagles to East Anglia. Andy gave me the go-ahead to write up an initial proposal, which he hoped would cover some of their reservoirs. I sent Andy this proposal on 23 March, leaving Tim Appleton and me to carry out a recce on part of the Suffolk coast on 4–5 April. Tim and I drove over in the rain, on a miserable day, to the Suffolk Wildlife Trust's headquarters outside Ipswich, where we met Julian Roughton and Alan Miller of the Trust, and also Andy Brown. I gave a presentation based on my report, and then we had a good discussion and reviewed maps of the coast. We had a very positive reaction from the Trust.

In the afternoon we drove north to the Blyth estuary, a fine wetland with open marshes, big reed beds, coastal fields and lovely old trees. It looked really good for sea eagles, with plenty of feeding and nesting sites. Alan took us by small side roads to Walberswick nature reserve and to Dunwich, where we could overlook the famous Minsmere RSPB reserve. The road to the nuclear power station gave us a view back up the coast to Minsmere, seeing red deer as well as a great deal of suitable habitat. Tim and I found a B&B in Aldeburgh and I enjoyed seeing a flock of avocets on the estuary before dinner. Early next morning we set off for Snape Marsh and looked down over the marshes to the Alde River, which was teeming with wading birds such as black-tailed godwits, curlews and dunlins. We continued by leafy back roads to Orford Ness pier, from where we could see Orford Ness itself and Havergate Island reserve, another superb area for sea eagles, and we discussed the idea with the RSPB warden. Next we headed inland to Ipswich, in order to meet Andy Brown at Alton Water reservoir, and we walked round. The water was busy with greylags, coot and great crested grebes, but I thought the place too busy with people. Mick Wright turned up and took us to view the Stour estuary, look over the River Deben and then see Trimley Marshes. The whole area was good for white-tailed eagles, with plentiful food and quiet loafing areas, including nesting sites in old woods. Over tea and cakes we discussed our day and decided we should take it further.

I rewrote my feasibility report for Anglian Water and sent it to Andy on 25 April. My view was that the Suffolk coast was an ideal area for the species and that this would be an extremely important and exciting project, with a high likelihood of success because of the quality of the coastal habitats and prey availability. I also mentioned that I had a recent telephone call from Phil Grice, the ornithologist with English Nature, known to us over the Rutland ospreys, who was totally in favour of the proposal and wished English Nature to be involved in any partnership.

My report covered the proposal in detail, including the rationale and the techniques, but in summary I was very impressed by the Suffolk coast, which I had not visited for about 20 years. It reminded me of northeast Germany and I thought something like 15 pairs of white-tailed eagles could live there. I was impressed that there was an extremely good mix of extensive coastal marshes, estuaries, rivers, reed beds, freshwater lakes and pools, rough grassland, grazing land, open sea, shingle beaches and islands, along with mature woods and open heathland. There were also plenty of quiet areas with dispersed farms, villages and towns. I was in no doubt that there was room for breeding sea eagles and could easily understand how the occasional wintering eagles in the past had been easily assimilated into this landscape. As I expected, there was high availability of live prey and carrion for eagles; potential prey species included feral geese – particularly the goslings – such as greylag, barnacle and Canada. These were not only good food for eagles but the predation might be viewed in a positive manner by farmers and golf course owners. There were also plenty of water birds like coot and cormorant, which are favoured prey, as well as a very large gull colony on Orford Ness. Large numbers of other wildfowl and waders also used the area. The large numbers of brown hares and rabbits, which are excellent food for eagles, particularly when feeding young, also impressed me. There was also the interesting possibility of their use of Muntjac deer as well as carcasses of red and fallow deer. The long stretches of open coastline would provide scavenging opportunities for the likes of dead fish, cetaceans and seals, as well as dead birds.

In our discussion in the office there was some concern that there might not be sufficient trees of a large enough size and structure for breeding eagles. However, during our field trip I had been able to ascertain that there were many opportunities for tree nesting, both in deciduous and coniferous trees, as eagles do elsewhere in Europe. There were also sufficient trees in quiet areas, as well as dotted over an open landscape. There was, in fact, no shortage of nesting sites on the Suffolk coast. The land owned by Anglian Water was not extensive enough to house a release location, but I was impressed by the areas around Walberswick, the River Blyth, Minsmere and the River Alde. I recommended two release locations; one in the north and the other about 20 kilometres to the south. It was very beneficial that there was a high degree of nature conservation ownership and management, with reserves belonging to English Nature, Suffolk Wildlife Trust, the RSPB and the Forestry Commission, as well as Anglian Water properties. I was encouraged to see strong working relationships between the organisations as well as enthusiastic staff on the ground. With regard to potential problems, the local farming systems did not suggest any issues, although pig rearing was mentioned, but these are busy places, eagles are shy and I had not heard of any problems with pigs in Europe. Despite the fact that this is a populated area in England, I believe it is very similar in many ways to the Baltic coast of Germany and that white-tailed eagles could be easily accommodated within the present land management. The rich prey base was better than in western Scotland, so we could expect higher productivity.

I was most encouraged by the visit and even more convinced that this area was excellent for reintroducing a breeding population of white-tailed eagles and that there were people keen to carry out the project. It would be necessary to have a strong partnership of a variety of interests, including farmers, estate owners, tourist boards, boat tour operators and other local interests. I also said that I was in contact with sea eagle workers in Germany and Poland, potential sources of young birds for reintroduction.

On 25 October 2005 I attended a meeting at the English Nature office in Bury St Edmunds, Suffolk, organised by Phil Grice and chaired by their area manager, Richard Rafe. We had a good group, including Julian Roughton from the Suffolk Wildlife Trust, Phil Grice and Richard Saunders from English Nature, Peter Newbury from RSPB, Adam Gretton, and Andy Warren from the Forestry Commission. I gave a talk and showed a PowerPoint presentation on white-tailed eagle reintroduction, which was followed by a wide-ranging discussion, including the feasibility of an English project, IUCN guidelines, public relations and partners; it was a very positive meeting, and by the time we stopped for lunch much had been agreed. In the afternoon I drove up to Thetford Forest to check its suitability, which I actually thought would be better for golden eagles, and then on to Wicken Fen National Nature Reserve, which did look possible. I then headed back to the Cotswolds to see Derek Gow release some beavers.

Following this meeting, English Nature started to put together their ideas and I assembled data on the history of the species in East Anglia and potential future breeding populations in England. At that time, I thought it possible that a future English population might reach 350–500 pairs; I now think that I was being conservative. I estimated 85 pairs for Norfolk, Suffolk and Essex. It was decided to hold a field meeting in early 2006, which took place in late January. The next day, when I was back home, I sent an email to Torsten Langgemach in Brandenburg, an old friend through our work with ospreys and peregrines, asking if he thought it might be possible to get donor young white-tailed eagles from northern Germany in 2007. His reply was cautious.

I returned to Bury St Edmunds on 24 January and met up with Adam Burrows and Richard Saunders of Natural England, and – from the RSPB – Alan Miller, Peter Newbury and Adam Rowlands, the site manager of Minsmere. We visited Benacre NNR, which I considered a very good release site, and then Hen Marshes, before going to the south shore of Blyth estuary, another good site; I enjoyed seeing a flock of 300 avocets. Dunwich was not suitable, but the Warren on Minsmere was the best release site, enhanced by the hordes of rabbits on the reserve. That evening it was great to meet up and stay with Cliff and June Waller at Walberswick; Cliff had been one of the assistant wardens at Fair Isle in the 1960s. The next day we went further south to Snape Maltings and on to Stanny Farm, where we met Paul Cooke. The south side of the Alde estuary was another excellent site; this was confirmed when we visited two Suffolk Wildlife Trust reserves on the north shore. It had been a really useful survey and, after discussion with Richard over a snack at the Maltings, I drove home to Scotland. I sent a report to Andy Brown of Anglian Water and thanked him for helping get the project underway and for funding my first visits. English Nature was clearly now running with the white-tailed eagle reintroduction and Richard was doing sterling work.

When I attended the osprey reintroduction workshop in Andalusia in late March, I had the chance to talk to Torsten about the possibility of getting young white-tailed eagles from northern Germany, but he did not think the people monitoring the species would wish to get involved. In consequence, I contacted another long-time raptor friend, Tadeusz Mizera, at the Agricultural University in Poznan; he was an expert on both ospreys and white-tailed eagles and he thought it would be possible to collect young eagles in Poland and transport them through the European Union to England. This was good news for Richard, and we started to plan a visit. Tadeusz collected us from Poznan airport on 18 October: he had organised accommodation for us in a small hunting lodge in the middle of the forest, belonging to Polish Forestry. Over a nice dinner, I had a very good discussion with him and the chief forester about the next day's activities.

It was a beautiful, clear cold morning, with a red sun rising through the mists in the old forest. A young forest ranger and Leszek Nowak arrived early with a 4x4 to take us to the first eyrie, in a big Scots pine. Later, Tadeusz arrived and we met Mr

Szelag, the chief forester, at his impressive headquarters within the protected forest. We discussed white-tailed eagle, black stork and crane, and there was no doubt that, with the necessary Polish government licences, this forest district would help to collect young eagles. We visited two more eyries before driving to the university in Poznan to meet Professor Andrzej Bereszynski. Reintroductions and translocations of great bustard, black grouse and hazel hen were his specialities, as well as wolves and European bison. He was very supportive of our request and, as chairman of the advisory group, which reports to the Polish Environment Minister, we asked for his advice on the best way to ask for licences. He gave us his contacts. In the evening, I gave a public lecture in the university on the restoration of white-tailed eagles in the UK to an audience of about 80 people.

Next day Tadeusz drove us to the forests west of Poznan, where he showed us several nests which had been successful in 2007. In the late afternoon we were at Poznan Zoo to meet the director, Radoslav Ratajszczak. He was experienced with reintroductions and captive breeding. Initially, he wanted to donate some captive-bred and rehabilitated eagles, but I explained that we would require wild young. After much discussion, they thought 15 young in one year was realistic, but it would call for a lot of very hard work from them; in an exceptional year they might possibly get 20. We left Poland very encouraged that they were keen to help us with donor young, which they would regard as a gift from Poland. We had discussed the work required to collect young, such as helpers, tree-climbers, timings and costs, and Tadeusz and Radoslav advised us to get an official letter from the head of Natural England to the Minister of the Environment in Warsaw as quickly as possible. It was looking good.

The proposal to release young sea eagles in Suffolk became public knowledge in the summer of 2007, and although many people were really supportive, and tourism providers saw it as a chance to spread the season for birdwatchers, opposition started to build from some farmers and landowners. It was not helped by sensationalist stories in the press. They would attack piglets! They would frighten the Christmas turkeys! An article by Robin Page in the *Telegraph Weekend* newspaper went so far as to depict an adult white-tailed eagle as the 'Bungay Beast'. Even our friends were a worry: one RSPB scientist suggested that if the sea eagles were released at Minsmere, the rare and EU-protected bittern would be at risk.

Natural England maintained their commitment to the project and they suggested, in view of the controversy in Suffolk, that we should examine the north Norfolk coast. On 30 April 2008 Richard collected me from Norwich airport and we visited the Broads national nature reserves, where a local warden gave us a tour by boat. All the habitats were suitable, but potential disturbance from tourist boats made it necessary to find quiet areas to establish a release site. The next day we were on the north Norfolk coast and visited an area where Richard knew the landowner was sympathetic. This was a beautiful estate on the north coast called Ken Hill, where we had a very good discussion and walk with the owner and the manager. It was an excellent place for a sea eagle release, with expansive views out to The Wash and the North Sea. We then drove east and met up with another landowner who showed us fine parkland and woods which would make another good release site, with excellent, ancient trees suitable for future nesting. We spent an hour at Titchwell nature reserve, looking at a great array of birds and talking with the wardens, before calling in on Aubrey Buxton. I had always wanted to meet him, because one of his ancestors organised the return of the capercaillie to Perthshire in the 19th century. Despite a family pedigree in reintroductions, he was sceptical about sea eagles. I think that, despite my reassurances, he was worried for the avocets he could see from his window. The next day Richard and I had a long drive to Essex to visit Horsey Island, a superb wildlife area and a potential site for eagles. I was thrilled when the warden showed me the nest of a shoveler in the long grass with 10 eggs, a first for me. On the way north we visited the RSPB's new reserve at Lakenheath, and I then stayed a few days near Marham with my son, who was serving in the RAF.

Despite Richard's continued good work on the project, our next meeting at the Natural England office in Norwich on 22 July 2009 seemed to throw up more questions than answers. I thought people were getting excessively worried about what could go wrong – electrocutions, lead poisoning and so on – instead of recognising the benefits. Later that month, though, Natural England approved the idea of two release sites in Suffolk and asked me to revisit Tadeusz in Poland. It was clear that some farmers and landowners were trying to slow anything down.

Duncan McNiven of Natural England and I went to Minsmere in the afternoon to look very carefully at the location we thought best for building the cages. The next morning, with the RSPB site manager, Adam Rowlands, we worked out exactly where the cages and the feeding nests could be built. The next day we drove all the way to North Kent to survey Elmley Marshes and Sheppey Island, a large area of grazing marshes and tidal flats, making ideal habitat for white-tailed eagles, although the nest sites needed to be inland. Michael Waters, the RSPB warden of Blean Woods, showed us through the wooded reserve, but – despite an extensive search – we could not find what I was looking for. The locations in East Anglia were better. Finally, we met the Natural England area officer at Stodmarsh, but for newly-flying young eagles, there were too many huge areas of reeds and bushes.

In 2010 we were still hoping that the project would go ahead, and I revisited Tadeusz in Poland to confirm that sea eagles would be available if the English project could be funded, with an agreed start. I was delighted to hear that they would be. Then big politics kicked in. On 11 May 2010 David Cameron was elected Prime Minister of a Conservative government, and someone with political clout persuaded Natural England to bury the sea eagle project. If only it had not taken so long to dot all the 'i's and cross all the 't's, I'm sure we would have had the sea eagles back by then. The lesson with projects is not to be slow; slowness allows opposition, often misjudged, to build momentum. We lost this one. The change of government also dealt a blow nearer to my home, when the new Nimrod surveillance aircraft – with a strong sea eagle connection – were scrapped and RAF Kinloss was closed at a combined reported cost of £3.5 billion. It's figures like that which clearly show that our society does not put enough money into wildlife conservation and global ecosystem protection.

## The possibilities for Wales

Several times, when I was in West and North Wales, looking at the potential for restoring breeding ospreys to the country, either by building artificial nests or by translocations, I also noted estuaries perfect for sea eagles surrounded by old woodlands with excellent nest trees. I remember the Dyfi and Mawddach estuaries as outstanding. In October 2003 I visited South Wales to walk up Sugar Loaf mountain, one of the highest peaks in the heart of the Black Mountains, where my father's ashes had been scattered on the slopes by my sisters some years before: it was close to where he spent his boyhood years. Afterwards, I walked parts of the Pembrokeshire coastal footpaths, near Marloes, looking out at the seabird observatories of Skokholm and Skomer Island, with Ramsay Island across St Bride's Bay to the north. Then I drove to overlook Castlemartin army ranges, where I had camped as an army cadet in the 1950s, and from the headland I could see really good areas for white-tailed eagles. The sand dunes and the great sweep of sandy beaches were ideal places for them to scavenge washed-up fish, cetaceans and birds. The next morning, I followed minor roads down to the side of the River Towy and its estuary; being low tide, the massive tidal flats were covered with waders, ducks and gulls, as well as some rod fishermen far out at the edge of the sea. From my vantage point I could look across to Pembrey Sands, another wild area of coastal habitats with pine forests behind the dunes. To me, it was prime white-tailed eagle habitat. In my view it was just a matter of time before someone restored sea eagles to Wales.

In 2004 the Countryside Council for Wales commissioned Mick Marquiss, a raptor scientist based at the Centre for Ecology and Hydrology in Aberdeenshire, to examine the potential for reintroducing golden eagle and white-tailed eagle to Wales. Mick's scoping study was delivered in February 2005. His conclusion was that, although both species could be restored to Wales, the case for white-tailed eagles was more compelling and likely to be more successful. The case for reintroduction was slowly moving forward, with Steve Watson, of Glaslyn ospreys, often pushing it along. On 3 May 2006 I attended a white-tailed eagle meeting at Bangor; on the way I met Steve at Portmadog. We viewed the Dwyryd estuary at Portmeirion, which I thought was excellent for sea eagles. The meeting, with about a dozen people, was at the agriculture college and chaired by Professor Gareth Edwards-Jones of the University of Wales, Bangor. The purpose was to try to start a reintroduction in Wales. Gareth opened the meeting, explained the purpose and asked for initial comments. I gave a presentation using PowerPoint about a potential translocation to Wales of white-tailed eagles and explained how we had carried out the projects successfully in Scotland. There was then a question-and-answer session. The most positive attitude was from the lady from the Welsh Tourist Board in North Wales, who was also the wife of a farmer, but the representatives of the RSPB and the Countryside Council for Wales were non-committal. I had a feeling the project could easily be kicked into the long grass, but Gareth organised a steering group to take it forward. Afterwards I drove over the Menai Bridge to Newborough Warren on Anglesey, and walked out through the pines to a high dune to view the reserve. It was yet another good location for a release site.

The project teamwas named Eryr Mor Cymru (Welsh White-tailed Eagle Reintroduction Group) and held its inaugural meeting on 1 October 2006, at which it was agreed that Bangor University would be the lead. They asked Mick and me to be consultants and needed some immediate help over a range of issues. In May 2007 Steve told me that they were getting support and that everything was going smoothly. He hoped I might come down in the summer to give a talk at a public launch, and I agreed. I was a bit concerned, though, that they were contemplating buying 120 satellite transmitters, which would be a major financial outlay, while the spreadsheet of work to be done was – while a work of art – quite daunting. Members of the group worked very hard on the feasibility study and supporting documents, including a 40-page first draft for discussion at their meeting of 12 February 2008. Apparently, it then stalled, waiting for someone, sometime, to seize the challenge.

Ten years later, that challenge of restoring eagles to Wales was finally taken up by a PhD student, Sophie-Lee Williams at Cardiff University. With her colleagues at the University and the Wildlife Trust Wales, she set up the 'Eagle Reintroduction Wales' (ERW) project to carry out an evidence-based assessment on the feasibility of reintroducing golden and white-tailed eagles to the modern Welsh landscape. Much consultation and research has already been carried out, and an interim report for the Welsh Government was produced in 2019. At last I have a real feeling that this will lead to the return of either or both eagles to Wales, and I am encouraged that it is led by Welsh ecologists with a strong knowledge of the Welsh language and local geography. I wish them all success.

## A sideways look at the Isle of Man

In November 2007, Nick Pinder, the general manager of the Curraghs Wildlife Park on the Isle of Man, wrote to me about his idea of promoting the reintroduction of sea eagles to the Isle of Man. We had met some years before at a reintroduction workshop in Hampshire. The wildlife park was administered by the Department of Tourism, and Nick thought that having sea eagles living on the island would boost visitor numbers, as well as restore an iconic species that had lived there long ago. I wrote to him with a positive but cautious reply, for I knew that there would be many hurdles to overcome. That, though, shouldn't stop a good project with a real champion to push it forward. I had never visited the Isle of Man, so I couldn't give much advice on its suitability.

This was put right when Nick invited me to visit him on the Isle of Man in 2009. On 21 May he met me at the airport and we did a series of excellent field visits all the way up the east coast of the island. There were more opportunities for eagles than I had expected, with quiet areas and suitable cliffs. I stayed in a B&B and, after a meal at his home, Nick and I went and walked at the Point of Ayres, looking north to Scotland. This was a wonderful area of raised beaches, pebble flats, dunes and heathland. A hundred ravens were flying around their roost in a pinewood and rabbits were very plentiful; I could easily imagine eagles hanging out there.

The Calf of Man off the south of the island was our destination next morning; it was a lovely half-hour boat journey away. I had heard about the bird observatory on this small island since my early years at Fair Isle; we were all part of the bird observatory family around the British Isles. Sarah Harris, the ranger, showed us around the bird ringing traps and we had a good explore of the island, even seeing an osprey flying north over the sea. There were lots of rabbits here too, and I could picture sea eagles visiting on hunting trips. In the boat back to the harbour, we passed close to two basking sharks. On our way back north, Nick showed me more areas of wild cliffs, bays and woods. In the evening, after supper, I had another walk on the Ayre nature reserve and listened to the nesting curlews while the ravens tumbled down to their roost. On my last day we checked out the inland moors, Glen Auldyn and out to Maughold Head; later we visited the hills near Sulby reservoir before an afternoon lecture on sea eagles to about 60 members at the AGM of the Manx Trust, while in the evening I repeated the talk for a meeting of the Curragh Friends at the Arts Centre.

It was a most interesting and enjoyable couple of days and, with Nick's guidance, I felt I had seen many parts of the Isle of

Man. There were areas where sea eagles could live and breed successfully, and there was sufficient food. My one hesitation was the relatively close proximity to the coasts of England, Wales, Scotland and Ireland. This might result in released young sea eagles leaving the island and not coming back; but, equally, when sea eagles are finally restored on those coasts, a release on the Isle of Man might encourage wanderers to the island to stay. Nick put great effort into working up a project in much detail, but, like other attempts before his, although he started to gain support, he also received negative and unsupportive responses. This time he was unable to make it happen, despite much well-reasoned argument. The Isle of Man is surely another place where sea eagles will – one day – return.

## South to southern Spain

In the early years of the new century, I had the great pleasure of working with Dr Miguel Ferrer and Eva Casado on their osprey reintroduction project in Andalusia. They worked at the Coto Doñana Biological Station in Seville, and I made many exciting field trips with the Osprey Project Team in those years. Occasionally, Miguel and I would talk about the progress of the white-tailed eagle reintroduction in Scotland, for he was also an expert on eagles, and would then dream of restoring the *pigargo* to Spain. On one occasion, we were on a field trip to the wetlands of Coto Doñana, an incredible wilderness for nature, and saw a flamingo with a damaged wing in a small flock walking across a lake. 'Excellent food for *pigargo*,' I said, and the others smiled. Another time, I was birdwatching in the incredible marshes of Isla Mayor, south of Seville, much of them now converted to rice growing. In a newly harvested paddy of rice covered in birds, I came across about 600 purple gallinules feeding on spent grain. The gallinule is a large, coot-like bird with brilliant blue plumage, large red bill, long legs and ungainly flight, and I could just imagine a white-tailed eagle sailing across the field, grabbing a gallinule and flying off to eat its exotically coloured meal.

In 2004 Miguel wanted to explore the idea further and asked me to give him some information on how we did it in Scotland, so I sent him a document of ideas and suggestions. I'd spoken to Duncan Halley, a Scottish friend working as a wildlife scientist in Trondheim, and he had reported that he and his colleague, Torgeir Nygård, were very keen to help. I recommended to Miguel that he should look seriously at the potential of reintroducing the white-tailed eagles to a place like the Coto Doñana, even though I recognised that the last records of breeding were from the north coast of Spain. I thought the potential food supplies and variety of prey species in that area might not be as great as in the areas with much higher food supplies in the southern marshlands.

I also felt that a very good case was to be made about the balance of species. In summer, the big wetlands of southern Europe no longer had a large natural scavenger and predator like the sea eagle, and the very large and undisturbed colonies of cormorants and herons – or, in Coto Doñana's case, flamingos and purple gallinules – were unnatural. The reintroduction of a large raptor might in fact be beneficial to the functioning of the ecosystem and for ecological purposes.

I offered, as I had the year before, to go to northern Spain to look at the potential for reintroducing white-tailed eagles, especially with regard to potential release sites, nesting and foraging areas. We now had great experience at reintroducing white-tailed eagles, gained over 35 years. The latest Scottish sea eagle newsletter gave the results for 2003; it had been an extremely good season, with a total of 26 young reared, double that of our best-ever year.

I reported to Miguel that we were in active discussions to set up new release sites in Scotland, with my preference being in the east, where prey availability was higher. I also sent ideas on the numbers of birds and all the techniques we used for the translocation, rearing and release of young sea eagles. I gave him contact details for the Norwegian ornithologists and a firm offer of help at any time. Later, Miguel asked Doriano Pando, a zoologist on the Bay of Biscay coast, to carry out an assessment; her report described the last couple of coastal eyrie locations from the previous centuries and included a note on present habitats and food availability. Other projects took precedence but Miguel told me recently that funding was looking likely for a reintroduction in Andalusia. It will be wonderful to see it happen.

## The Rock of Gibraltar

After a Raptor Research Foundation conference in Seville in September 2001, some Spanish friends drove me down to the Barbate reservoir, near Alcala, where ospreys had been occurring with increasing frequency. We visited the local nature reserve near Algeciras, where, for the first time, I met John Cortes, the famous ornithologist from Gibraltar. He wanted to know about the satellite transmitters which we were attaching to ospreys to study migrations from Scotland to West Africa. He was very taken by the results and wanted to do the same with some of the birds that he studied. We had a good time together, watching birds in the estuary, and I took him back to the border so that he could walk across to Gibraltar. One day, he said, you must come and visit our bird observatory on the Rock.

This finally happened on 13 December 2005, when John asked me to visit Gibraltar to give my expert views on the potential for restoring breeding ospreys to the Rock, where they once nested. He had been watching closely the reintroduction of ospreys being carried out by Miguel Ferrer and Eva Casado at the Barbate reservoir. I drove to the border at midday from Ronda, John met me, and I followed him by car to the bird observatory accommodation at Bruce's Farm, with a great view over the Bay of Algeciras. I dumped my bags and car, and John took me for an exploration, first at the south end, then round past the rubbish dump to view breeding peregrines, and he showed me

The cliffs of Gibraltar – an ancient sea eagle nesting site.

the old site where ospreys nested. There were huge numbers of mullet feeding at the sewage outfall. Then we travelled up the eastern side of the Rock, with its fantastic high cliffs, and back through the town to the Botanic Gardens for six o'clock. They had agreed I should give a seminar on the reintroduction of ospreys, red kites and white-tailed eagles to a small group of zoologists, and we had the usual great discussion, with a lot of interest shown.

The next morning I was collected at 9.30 am by Paul Rocca and Vincent Robba, who wanted to show me peregrines. They had seven pairs nesting on the Rock, and recently a pair of eagle owls. They took me to their Raptor Centre at the southern end, where they had a small selection of raptors, and where they also look after raptors which get into problems crossing the street. They told me about vultures being forced to sea level by a mob of angry yellow-legged gulls; they sometimes rescued them by boat and rested them for a day or two. I looked at the centre's potential as a release area for white-tailed eagles before we returned for lunch. That afternoon we walked to the top of the Rock, where tourism is rampant and the apes cause a lot of havoc. The next morning John collected us and we had a most interesting day, during which the security guards and cavers took us into the amazing network of tunnels deep inside the rock. The large holes in the sheer cliffs, which had originally housed big guns, would make great places to build artificial nests for sea eagles.

In the afternoon John took me to the Environment Department and we met Jaime Netto, the Minister. We had a very good reception and discussion, with a very positive 'Yes!' for John and his team to start a sea eagle project. We later went to another part of the summit, which is a prohibited military area, where they opened the locked gates and we could view the whole of the east cliffs from halfway up. There were many safe roosting sites. That evening I gave a talk in the town to 30 people about my work in Scotland and again took many questions, which continued when they took us to the Fishermen's Club on the seafront for a fantastic seafood meal. One of the most fascinating bits of information I learned was that during excavations in one of the caves in the eastern face of the Rock, archaeologists had discovered sea eagle bones in association with the remains of Neanderthals. This is just like it was in Orkney, where ancient humans had been buried with the bones and claws of white-tailed eagles. This great bird had a special significance for our ancestors. The next morning I left to meet osprey friends at the Barbate reservoir.

During my visit to the Rock I was most impressed by the opportunities that I saw for using Gibraltar as a focus for restoring the white-tailed eagle to the western Mediterranean. When I got home I wrote to John with a full assessment of my visit; I had to say that I did not think there was merit in releasing ospreys at Gibraltar, because the Spanish release site in Andalusia was only 30 kilometres away. If that project was successful, then building artificial nests on the Rock would be a good idea. I did, however, enthusiastically support a project to translocate and release young sea eagles on Gibraltar. Just a couple of pairs might breed there, but it could be a catalyst for a larger potential breeding population in the western Mediterranean; nest sites in southern Spain and North Africa might mean that a population of maybe up to 10–15 pairs could be established, if more than 60 birds were released. Of course, even if a few pairs started to nest, there would be moves to release more birds in Spain, France and Italy. I even thought they could recolonise the North African coasts. The Gibraltar project could be the launchpad for a larger reintroduction.

I gave John a detailed account of how to run such a project, covering all the issues that we had learned in our reintroductions. I said we would help and provide him with all the necessary contacts and techniques. They clearly had the right conditions for a project in Gibraltar, but it needed organisation and it needed funding.

After that, things went quiet – which is not unusual with reintroduction projects.

Then, out of the blue, in April 2014, I got a phone call from John Cortes, with the news that he was now a politician. Not only that: he was the newly appointed Minister for Health and the Environment of HM Government of Gibraltar, and sea eagles were back on his agenda. I congratulated him, said we were ready and, by email, added, 'Now I just need to be voted Environment Minister in Scotland, and I could reintroduce the lynx!'

Stephen Warr from John's office became our main contact, and I said that I would visit Gibraltar at the first opportunity. On 11 June I was collected by taxi from the border gate and taken to the Jew's Gate bird observatory. Stephen soon arrived and he took me up to the north summit of the Rock, where the military had relinquished some land. The old buildings would make an ideal release site for eagles, and there was no doubt that his colleague, Reuben Senior, could undertake the work. I had been impressed by the amount of scrub clearance that he and his team had done on the slopes for the conservation of Barbary partridges. They dropped me at a meeting of the Gibraltar Ornithological and Natural History Society in the town; after the finish of their business, I gave an update of the latest sea eagle news in Europe.

Soon after dawn, I was having my breakfast on the terrace, looking out over the Strait; the only bird I saw, apart from gulls, was a blackbird. At eight o'clock we were in the Minister's office, discussing the project with John and his environment team. It was a good meeting. They were keen to proceed and I said I would dust down the feasibility reports and renew contact with the Norwegians. I also suggested we team up with Miguel and the Migres Foundation, for the eagles would quickly visit Andalusia. The Minister's nature conservation committee were still apprehensive about predation of the small number of shags nesting in the caves, which belonged to the Mediterranean subspecies. Duncan and I assured them that the sea eagles would not go into caves to catch shags. That concern, though, was something which unfortunately didn't go away. After a visit to the south point I was taken back to the airport to catch the plane to London.

With the help of Duncan and Miguel, I brought the feasibility report up to date and also sent Stephen a schedule of advice and pertinent dates for his team, so that everything would be ready when the birds arrived in 2015. I returned to the Rock on 13 October and was soon installed at the bird observatory, where the bird ringers were busy catching and ringing birds in the mist nets. I enjoyed seeing Sardinian warbler in the hand, a blue rock thrush on the roof, and they also caught a Scop's owl and nightjar. In the evening, Duncan arrived from Norway. The next day, Stephen and Reuben collected us, and when Miguel arrived we drove to the summit of the northern top of the Rock, to show the others how ideal it was for rearing and releasing the eagles. At midday we had a meeting with John's environment team to discuss the IUCN feasibility report, as well as Miguel's report on scientific monitoring. Duncan was sure that the Norwegians could supply 20 young a year for five years.

During the following months we completed an update of the main document; Duncan recommended that the annual numbers should be increased to 20 young per year and that we should draw attention to the survival of griffon vultures in the nearby Spanish mountains. The sea eagles were likely to feed with the vultures, and the lack of poisoning reports was important for their survival. We discussed a visit in the spring for me to organise training the team to build the infrastructure, and I had also spoken with a wealthy friend who offered free transport for the birds in a private plane from Norway to Gibraltar. Unfortunately, as so often, the project ground to a halt, showing again how difficult these bold proposals can be. I hope, one day, that white-tailed eagles will be restored to the southwestern Mediterranean.

## Exploring opportunities in France

At the invitation of Jean-Marc Thiollay and Yvan Tariel, Mission Rapaces of LPO (French Birdlife), I was asked to give my expert view on the suitability of the Champagne lakes, in northeast France, for the reintroduction of white-tailed eagles, and to look for potential release sites. I arrived at Troyes on the evening of Tuesday 13 November, which would give me two full days in the field with Jean-Marc and Rolf Wahl, an old friend from Orléans. On 14 November we visited Lac du Der-Chantecoq and the Etang des Landes, where we walked through the reserve to two hides; then we went to the southeast corner of the lake and walked to look at several parts of the Presqu'ile de Champaubert. In the afternoon we drove to the Etang de la Horre, where we met reserve staff who guided us until dusk. The next morning we went to the Forêt d'Orient park research office and travelled with the warden, Stéphane Gaillard. He showed us many parts of the shoreline of Lac du Temple and

Lac Amance, as well as a line of other small fishing lakes in the forest. We visited the south end of Lac d'Orient and travelled up the eastern side, back to his office. At lunch we checked the north shore and then returned to the park HQ. In the afternoon, the Director, Thierry Tournebize, joined us and he guided us through the reserve, with Stéphane Gaillard, to look at various habitats and locations down the peninsula in Lac du Temple.

The two big lakes and their associated smaller lakes and woodlands were, in my view, very suitable for a breeding population of white-tailed eagles. I estimated that a population of at least 10 pairs, and possibly up to 20 pairs, could live in this region. The local farming system had a couple of advantages: there was no potential conflict with sheep breeding, and the extensive land areas without habitation would allow eagles to rest without disturbance and possibly find carrion. Ecologically the return of an apex predator which would prey on cormorants and other abundant species could be beneficial. There was no shortage of potential food for sea eagle. There were very large numbers of a wide variety of waterfowl, including ducks, geese, cormorants, herons, grebes, coots, cranes and gulls, on which they could prey, and with such large numbers of birds present there would be weak, dying and injured birds which the eagles could kill, as well as carrion. It would be necessary to check the availability of birds for prey in the breeding season. I was informed there were very good stocks of fish in lakes ranging in size from small to very large, with favourite food species for white-tailed eagle, such as bream and carp. The eagles could also steal fish from cormorants and other species, so this would also be a good food resource. On a short visit it was more difficult to judge the availability of mammal prey or the extent of carrion. There would appear to be high levels of red and roe deer, and wild boar, so some natural carcasses would be available through the year; and there would be the opportunity to put out carcasses or part carcasses of these mammals, either from management control within the reserves or the transfer of road kills to future eagle-feeding sites. I was also struck by the numbers of non-native muskrat and coypu, which were likely to be favoured prey for white-tailed eagles. Trees suitable for nests were scattered throughout the region. In the forests of Lac du Temple and Lac d'Orient there were large and suitable oaks and beech trees.

During our guided visits I found several good sites to build release cages, and there was no doubt in my mind that there were competent staff to carry out the project. At the end of my brief visit I considered this region to be good for white-tailed eagles and felt it could be an ideal site for carrying out a reintroduction. It was a great advantage that it was reserve land, under the control of the Natural Park, which could supply necessary equipment, occasional manpower and possibly carrion.

Unfortunately, no further action on reintroduction took place, but it was very interesting to hear in 2011 that a pair of white-tailed eagles started to breed in the region. It's very important to remember that France is a big country: the real breakthrough would be to reintroduce the *pygargue* to the Gironde estuary and further south and east in the Camargue.

## Trying to restore sea eagles to Mallorca

On a visit to the Alladale wilderness reserve in Sutherland, I met a Norwegian, LC Kvaal, who took an interest in my conservation work. In autumn 2012 I visited him on his island in Norway to give advice on ecological restoration and, with his girlfriend Eliza, we built a big nest in a cliff to try to attract sea eagles. He told me he also lived on Ibiza in the Balearic Islands and, a year later, he said he would like me to visit and give advice on the potential of reintroducing white-tailed eagles to Mallorca. He was friendly with the people who ran the Black Vulture Foundation there and wanted me to meet them. We were still failing to restore the species to the western Mediterranean, where the last survivors were in Corsica in the 1950s. There were rare records of wandering juveniles reaching southern Europe in winter, including an immature sea eagle on Mallorca in the winters of 2000, 2001 and 2002.

At the beginning of April 2013 I went to Mallorca to meet LC and explore the possibilities. I made my way to a mountain village below the high cliffs of northern Mallorca, where I was booked into a hotel among the olive groves. LC and Eliza soon arrived and we went to meet Juan José Sánchez Artés, the black vulture expert, who drove us up into the mountains. We walked to a ruined castle to look out over a whole series of valleys and great cliffs, with a small group of black vultures hanging above us. Far in the distance was the Mediterranean Sea. To me, it looked good for sea eagles, and over dinner we made a plan. The next morning we drove to an incredible hidden valley called Ariant, back-dropped by the Tramuntana range of mountains. In the 4x4 we rattled on down towards the coast, from where I counted 500 Balearic shearwaters, 10 Cory's shearwaters and other local birds, three bottlenose dolphins. It was a beautiful, magical place, and with distant black vultures drifting along the ridges, I could easily see white-tailed eagles living in this place. For me the sight of hoopoes and woodchat shrikes was a reminder that I was in thc Mediterranean.

In the afternoon I was taken to the famous Albufera marshes, which were much bigger than I expected; there was a scattering of ducks, including red-crested pochards, as well as a few crested

The north cliffs of Mallorca – excellent sea eagle habitat.

coot and numbers of waders, but it was difficult to see what was really living in the marshes because they were so extensive. The next day I visited the Black Vulture Foundation centre, where I met Juan's wife Evelyn, and we talked about vulture conservation, the Ariant mountain estate and the idea of white-tailed eagles, but I felt they already had enough on their plate without an extra project. Juan kindly offered to drive me out to the northeast corner of the island so I could check more of the habitats and, again, I was convinced that sea eagles could be successfully reintroduced.

When I got home I prepared my report and impressions to send to LC. There was no doubt that restoration to the Mediterranean Basin would be a strong conservation initiative and that it would re-establish a key species in the marine and wetlands ecosystems. The evidence from reintroductions in the British Isles (and active restoration of bald eagle populations in the US by the hacking of juveniles) indicated that restoration would be successful in the Mediterranean Basin. I noted that the Tramuntana range held large areas of suitable nesting habitat for white-tailed eagles. The mountains rose to 1,200 metres and there were sea cliffs of 400 metres in height. The landscape was very broken, with mountains, cliffs, crags, open land, valleys of shrub and woodlands, as well as – in places – farms and cultivated land. The potential nest sites included cliffs, vegetated ledges and large trees. White-tailed eagles are generalist feeders capable of feeding on carrion of many forms, as well as fish, birds and small mammals that they catch. As juveniles and pre-breeding adults, they would most likely feed on carrion and probably associate with the black vultures and ravens at chance kills. The fact that the vulture group had arrangements and EU derogations regarding the disposal of dead farm animals meant that carrion from this source was readily available for carrion eaters. There was also a large population of feral goats in the mountains, and while many of these animals may be hidden in caves, rocks and woods when they naturally die, chance carrion would be available. The white-tailed eagles have an advantage over the vultures in that they can pick floating carrion, dead fish and birds, from the surface of the sea and will also patrol beaches for carrion such as fish, cetaceans, birds and mammals. As regards catching of live prey, which is important for breeding adults feeding young in the nest, they are likely to catch rabbits with myxomatosis, and seabirds such as young yellow-legged gulls and injured individuals. They are likely to visit wetlands like Albufera to predate wildfowl and coot, particularly sick and injured individuals. Non-native carp in the channels at Albufera could also be suitable prey, as would shoaling mullet close inshore. The fact that Mallorca hosts well over 100 black vultures, which breed successfully, as well as a newly arrived breeding population of griffon vultures, indicates that a small population of sea eagles would find sufficient food to breed successfully.

Some worries were expressed about their impact on rare birds such as Audouin's gull, Balearic shearwater and several water birds at Albufera such as crested coot, purple gallinule and marbled teal. Sea eagles live in areas with shearwater colonies – Manx shearwater at Rum and Canna in western Scotland, for example – but as the shearwaters come ashore at night, there have not been predation concerns. In Mallorca the night period is longer than in Scotland, so predation would be highly unlikely. Audouin's gulls are probably very demonstrative at the breeding colonies and can probably deter attacks. The water birds at Albufera are on relatively small water bodies and canals, so have plenty of space to avoid eagles unless they are in poor health or injured. The sea eagle is a generalist that cleans up sick and injured birds within populations. It cannot catch agile smaller birds such as partridges and pigeons, so would not conflict with hunters, while predation of rabbits and hares would be of sick and injured individuals. If white-tailed eagles were established on Mallorca, I thought they could regularly visit or settle on any of the islands in the Balearics, and that some could disperse to other parts of the Mediterranean Sea.

My report included a full outline of the hacking and release techniques, the potential partners and an estimate of costs, and

I suggested the release of 20 birds per year for three years. I concluded that a reintroduction of sea eagles to Mallorca and the western Mediterranean Sea was highly likely to be successful and would involve the restoration of a major species to the marine and wetland ecosystems. In the Balearic Islands it could result in as many as six to ten breeding pairs, with other released birds settling in adjacent regions. The best scenario might be to bring together a small group of keen and competent island and mainland Spanish ornithologists, plus representatives of Gibraltar and Corsica, to discuss the proposal and work out a united programme of releases in other locations.

LC subsequently met and discussed the idea with a range of island ornithologists and island officials, but although some were in favour, others were hesitant, and the proposal remains in abeyance. He worked diligently, through to 2018, trying to persuade the authorities to give the go-ahead, but other projects in other parts of Spain and South America were claiming his attention, so this one was laid aside for the time being.

This chapter demonstrates that for every successful project that gets the go-ahead, there are others into which people put great effort but which then run up against problems. These can be very varied: human nature comes into it, timing can be wrong. As the biologist told me back in Whitehorse, it's really important to have more than one project running at any one time: when one stalls, you concentrate on the others. I'm always prepared, though, to take a project back out, dust it down and give it another crack – it's the only way that progress will ever be made.

An aerial view of the Needles, Isle of Wight – looking east at potential sea eagle breeding habitats. (*Photo by The National Trust Photolibrary/Alamy Stock Photo*)

11

# LEGACY

## *New projects with the next generation*

This last chapter is in some ways about a beginning, rather than an ending. For me, it's about handing on the baton. I can no longer climb tall trees to ring young ospreys, nor scamper up to the high tops of the Cairngorms at dawn and be back for a late breakfast. I still have a desire, though, as great as ever, to push the recovery of birds and mammals and to work for much greater ecological restoration. I enjoy helping others, younger than myself, to carry forward projects that are important to them and to me.

I was fortunate that I met Tim Mackrill when he was a teenager volunteering with the Rutland ospreys. When he was at university in East Anglia he came north in the summer of 2002 and stayed with us, completing a six-week dissertation on fishing ospreys with Fiona Macphie, who was with the Foundation for a few years after gaining her doctorate. When Tim was in charge of breeding ospreys at Rutland Water, working with the Leicestershire and Rutland Wildlife Trust, we often met to talk about new pairs and to plan the satellite tracking of those ospreys. We often got round to talking about all the satellite tracking data sitting in my computer, and this led to Tim using that data to complete a PhD on osprey migration.

In 2016, when he was approaching the end of his research, he asked if he could join me and work with our projects. I said I would be delighted but that he would have to help look for funds, because the Highland Foundation for Wildlife had very limited resources. I also said that I would change our name – for 'Highland' was not a helpful name for our work on projects in the south, or overseas. I spoke with my trustees and we agreed to change the name to the Roy Dennis Wildlife Foundation, which we thought would make it easier for Tim, and others, to raise funds for future projects outside Scotland. This was a brilliant move for me: not only could Tim come up to Moray to

Sea eagle team at the Isle of Wight site checking young eagles prior to release. From left to right, Steve Egerton-Read, Tim Mackrill, Ian Perks and author, August 2019.

assist with osprey studies and the translocation of red squirrels, but he was on the spot for our southern projects. These were soon underway, with the restoration of breeding ospreys to the south coast of England. We chose to do it in Poole Harbour, working with a very strong local NGO, Birds of Poole Harbour. Tim also took over the research advice side of the Sussex White Stork recovery project, which I had supported strongly with Derek Gow and Charlie Burrell. Much of the planning and execution of our latest project – the recovery of breeding white-tailed eagles to the Isle of Wight – was done by Tim. He is also a friend and colleague to others who help us with projects in the north, like expert tree climbing conservationists Ian Perks and Fraser Cormack, while others dip in and out of projects as great friends and co-workers.

## The south coast osprey project

Whenever I visited my parents in Hampshire and went back to my old boyhood haunts, I would think that they looked right for breeding ospreys. I had always loved the old Hampshire name, the mullet hawk, and so the idea of seeing the birds in that landscape had been in my mind for the whole of my life. On 20 April 1993 I called at the English Nature office in Lyndhurst to visit Colin and Jenni Tubbs. Colin was one of my birdwatching pals in the 1950s and had done sterling work for nature conservation in southern Hampshire for 30 years. It was good to catch up with news of the New Forest, but I particularly wanted to discuss the possibility of reintroducing ospreys to the estuaries of the Solent. He liked the idea and we agreed to talk again. I revisited a few coastal sites, which I remembered from my birding childhood, but in the face of general antagonism towards reintroductions from the official bodies at that time, I left the idea on the back burner.

Colin's death from cancer just four years later was a tremendous loss for nature. In late October 2000, when I was south to see my mum and later to attend a Rutland Water osprey meeting, I spent a couple of days exploring the south coast estuaries all the way from Poole Harbour to Cornwall, for they reminded me of an earlier trip to the breeding ospreys of the New England coast in North America.

In November 2003, on a family visit to Hampshire, I called to see another old friend, Barry Duffin, the warden at Titchfield Haven nature reserve, and talked to him about – among other things – putting up an osprey nest. On my way back to Gatwick airport the next day I had a chance to look at various reservoirs in Sussex and thought they, too, would provide good habitat.

In May 2008 my wife Moira and I were in New England with my friend Alan Poole, a world expert on ospreys. After days of fieldwork, I gave three evening talks, at Newport, Gardner's Island and Nantucket – all great places for ospreys – and met a terrific bunch of osprey enthusiasts who were highly skilled in building nest poles in the estuaries. Some places, well stocked as they were with fish, resembled the south coast estuaries of England, with their great quantities of grey mullet.

In August that year I was promoting my book, *A Life of Ospreys*, at the annual Bird Fair at Rutland Water. I'm not a regular visitor, so it was great to catch up with many friends, old and new, and one particular discussion led to some funding to encourage me really to explore the south coast option. By great good fortune, I was asked to give an osprey talk at the annual conference of the Hampshire Ornithological Society in Winchester in March 2009. There was a big audience and a great deal of interest in encouraging ospreys to breed again in

Hampshire. As a starter, I recommended the building of about 50 osprey nests along the Hampshire coast and in adjacent parts of Dorset and Sussex.

The following September I carried out a field survey with Barry Collins and Pete Potts, starting at Chichester Harbour, and found plenty of suitable places, both there and on Hayling Island. We then visited the River Hamble, the backdrop to my childhood, and the following day, with Manuel Hinge, we had a great time around the New Forest, on the River Test, the Beaulieu River and the Solent coast as well as the River Avon. We had lunch in a pub near Blashford Lakes where we met the warden Bob Chapman and the Forestry Commission head keeper Andy Page. There were so many places that would be suitable for ospreys, and Manny turned out to be an expert nest builder. I had a similarly encouraging time the next day at Poole Harbour, where Mark Singleton of the RSPB and Jason Fathers arranged a great tour. There were so many excellent sites there: it was prime osprey habitat, with masses of grey mullet.

Back home, I got out my maps and, working with Google Earth, tried to envisage the potential in those counties, coming up with a suggested tally of between 62 and 84 suitable sites for artificial osprey nests to encourage a population of breeding ospreys. Manny and I moved things on, and on 10 December we organised an osprey workshop at Blashford Lakes Visitor Centre, attended by a good number of local wardens and birders. Tim Mackrill came down from Rutland Water, and in the afternoon we gave a demonstration of nest building. The next day Tim and I were back at the Arne RSPB nature reserve in Poole Harbour to give a demonstration of osprey nest building with reserve manager, Dante Munns, and Jason; we cut the top off a big Scots pine and, between us, built an excellent nest from local materials. We took our time, but it was important to get the techniques correct, constructing a really big, tall nest, with the correct lining of leaf mould and a nice perch. When we had finished, we were invited to the RSPB Christmas lunch in Wareham, where I enjoyed meeting up with two old RSPB colleagues, Brian Pickess and John Day.

The following April I returned to Poole Harbour and thought that the three nests that had been built on the Arne reserve were looking really good. The RSPB press officer had brought along a crew to film a piece for breakfast TV, and at the best nest we put two decoy osprey, which I had made from polystyrene. The perched one looked so good by the nest that it convinced a passing adult herring gull to start squawking and it came over to mob it. During the afternoon, Mike Dilger came to present a television report for *Inside Out*, so the idea of getting ospreys back to the south coast was rapidly gaining ground.

The next year I was on the Isle of Wight, giving advice to John Wilmott about putting some nests near Newtown Harbour, and made other visits to Cornwall, Devon and Somerset. In November 2015 I was giving a talk at a mammals conference in Winchester and the next day had a chance to go to Poole Harbour, where, as well as meeting Jason, I met Paul Morton of Birds of Poole Harbour, set up with local resident Mark Constantine. We had a really good talk and drive round the harbour. The nests had been up for some years and I was convinced that we should move to a translocation. I had been favouring the Beaulieu River estuary, but after our day together my thoughts turned to Poole.

Tim was in Scotland with me in July 2016 when we were collecting 12 young ospreys for the translocation project in Spain. It gave us a chance to talk more about the south coast ideas. Tim had been involved with the Rutland Water project and had watched the population slowly grow and be successful,

Author painting dummy polystyrene ospreys as decoys at Rutland Water in 1989. Similar decoys were used at Poole Harbour in 2011.

and had also made visits to the Welsh ospreys, the population started by two males translocated to Rutland Water. The Spanish project was coming to an end in 2017 and this seemed a good time to start a new one at Poole, on which Tim could take the lead with Paul Morton of Birds of Poole Harbour. That September we held a really good meeting at the office in Poole, and Paul was able to say that there were sufficient funds to start the project. With Paul, Jason and Tim, I had a really enjoyable search for a good release area and for a site on which release cages could be built. We decided that there was no time to spare, so we all helped to produce a feasibility report, based on previous documents that I had prepared for other reintroductions. We agreed to call it the Poole Harbour Osprey Project, and that it would be a partnership to establish a sustainable breeding population of ospreys at Poole Harbour and along the south coast of Dorset and Hampshire, run by Birds of Poole Harbour,

Wildlife Windows and the Roy Dennis Wildlife Foundation. Our aims were to re-establish a sustainable breeding population in Dorset by translocating and releasing young ospreys from Scotland at Poole Harbour, in order to increase the breeding population in the British Isles and to restore the species to its previous range, where practical; to encourage a larger geographical spread of the species by establishing a breeding population on the south coast of England and thus enhance the species' long-term chances of survival; to encourage breeding ospreys to use the rich estuarine habitats in southern England; and to create potential links between the English osprey population and breeding ospreys in France.

The rest of the autumn and winter was taken up with talking. We had meetings with Natural England and the local nature conservation bodies in the area, and with landowners, fishing interests and a range of other organisations who might wish to comment on the proposal. Forestry Commission England and Natural England were very supportive, as were various local landowners, especially as we wanted to use quiet, private land to establish the release location, in order not to disturb newly fledged ospreys. Jason was tasked with building the hacking cages to our well-tested design, and with ordering and installing CCTV cameras for observing the young. Paul and Tim dealt with meetings, and Tim led the production of the final feasibility report for Natural England and the application to release ospreys. Given that I live in the north of Scotland, I was delighted that Tim could take on the southern part of the project, as he lived much closer, while in Scotland I applied for a licence to collect and translocate up to 12 young ospreys per year for five years.

Late May 2017 was our final planning stage, and Tim and I went down to Poole for two days. We approved the choice of release site and staked out where exactly the cages should be built, as well as settling on a location for the team caravan, both for the CCTV monitors and as a base for staff and volunteers on duty. We talked with Brian Cresswell about Biotrack tail transmitters and visited Jason's workshop to see the cages; his workmanship was second to none. That evening we all had an enjoyable meal and great discussions in a fish restaurant, hosted by Mark and Mo Constantine, who wished the project all success.

In May and June I carried out the monitoring of all my regular nests in the north of Scotland, so that by the last days of June I had worked out which were the most suitable nests for collecting young for translocation. This was a more than usually busy year for us, because we first needed to finish the final translocation of 12 young for the Basque Country osprey project, and then turn our attention to Dorset. Aitor Galarza left my home at 4 am with 12 young ospreys, bound for Aberdeen airport, Emily Joáchim going with him to feed the birds before they took their flight to London and on to Spain. (Emily was our expert feeder of all the young ones while we were out collecting more in the forests.) Tim and Emily then set off by transit van for Poole with eight more young, which had just been given a good feed of fresh trout and were lying on moss in individual boxes. They stopped the night with friends of Tim's at their home near Stafford, giving them an opportunity to feed the birds and have a rest before a dawn start for the last stage of the journey to Poole Harbour.

Brittany Maxted, a local from a nearby village who was studying at Oxford University, was taken on to run a project on the ground with the volunteers. They started with eight young so that they could learn the exacting routines with a smaller number than the future cohorts of 12. Everything went to plan and, on 30 July, Tim collected me from Gatwick airport and we drove to Poole, where we met the team: Paul, Jason, Brittany and some of the volunteers. We also met Ruth Peacey, who was starting to record a film of the osprey project over the five years. My first look in the cages showed me that they had done a great job in rearing the young to fledging: at least six were ready to be released. During the day we caught all the young ones to check their weights and measurements, and to fit tiny VHF radio transmitters to their central tail feathers. These act as an insurance policy so that the birds can be found in the first few days of freedom – an inexperienced youngster might land in long vegetation and become trapped, or be mobbed by a large bird and get stuck in a bushy tree.

The team were at the cages at 4 am the next day, before dawn, and the first two cages were carefully opened 45 minutes later. It was decided that six would be released, with the two smallest ones kept back until 8 August. The first bird flew out at 7 am for its inaugural flight and within a minute was mobbed by a passing peregrine falcon – what an ordeal for its first outing – but it flicked in the air like a professional. Slowly, the others came out and flew free. It was superb to be sitting in the early sun on a bank, on the other side of the bay, watching the birds coming in to use the T-perches which had been installed for them on the edge of the field and in the marsh. The first ones then flew to the feeding platforms in front of the cages to eat fresh fish. It was so exciting to see them ranging on the edge of Poole Harbour, and one bird even flew across the water to perch in the marsh at Arne. Another got caught in willow bushes but was retrieved and placed back in a cage until the following morning. Later that day, when I got home, I learned that all the other five had returned to the cages for food, and soon all the young ones were flying well and had started the month-long period of learning how to be an osprey in the world. Brittany and her team made certain that they had plenty of fish whenever they returned to the release site. Whenever something interesting happened, I would get an email or a phone call to bring me up to date. It was very exciting when I learned that a Rutland Water-ringed two-year-old female, Blue CJ7, had turned up and interacted with the young. To me, this was really significant. Between 25 August and 13 September all eight young set off on migration for Africa, making a great start to the project.

During the winter we received information that two of the young ospreys had been identified in the Sine Saloum National Park in Senegal. This is an absolutely fantastic wintering place for European ospreys, where both Tim and I had been in winter and seen colour-ringed ospreys from the UK and from mainland Europe. A third youngster was later found dying in Gambia with injuries which suggested that it might have been grabbed by a crocodile, for the tail and back part of the body was missing. We thought it excellent that at least three of the eight young had migrated successfully to the correct region of West Africa.

On 12 April 2018 Tim and I spent a day at Poole Harbour, first going to the release site. We were pleased to see that more of the gorse bushes surrounding the cages had been removed, which would give the young much less chance of becoming grounded. We discussed where the two extra cages should be built, and suggested that seven more T-perches should be placed in the vicinity. We called by for a chat with the landowner and then went to the RSPB reserve at Arne, where an osprey had been seen at one of the nests. We walked up the hill and looked down towards the water, and I saw a female osprey catch and carry a flounder to one of the best nests, where it started to eat. My colleagues with their telescopes confirmed that it was blue-ringed CJ7 from last year. This was her third year; if only she could find a mate, she was capable of breeding. Then we saw another osprey fly to a different nest, but it was another female. CJ7 persevered and stayed all summer, so she must have been persuaded by the fledglings last summer that this was an osprey breeding locality. We went on and looked at other nests on Forestry Commission land and around the harbour, and finally called to see the new Birds of Poole Harbour headquarters on the Quay. The project was continuing to be a really exciting venture.

On 4 July Tim arrived in Scotland and helped me construct the pens for the ospreys in my garage. We discussed which nests were available, as I had checked all of them for young in the previous weeks. Emily arrived to carry out the feeding of the young, and on 7 July we set out for a couple of the nests but managed to collect only one young. The next day was much better. We spent it with Alan Campbell, the wildlife ranger with Forestry Commission Scotland in Moray, and collected six young in the morning, with more that afternoon and evening from nests nearer to my home. On 10 July Brittany and I went for fresh trout in Aviemore, and then met the climbing team in Badenoch, where we collected the final chick. That all sounds pretty simple, but it's not. Many of the trees are really high and difficult to climb, the tallest being nearly 40 metres, so we need a really great climbing team. Fraser Cormack and Ian Perks are highly skilled and capable tree climbers with the best equipment. In turns, one of them will climb up to the nest, place the young in a big, safe bag and lower them to the ground. We gently take out the young ospreys, weigh and measure them, record in my notebook the age, sex and condition, and then ring and colour ring them with unique combinations. As we do this, we are checking to see if there's a chick suitable for translocation: it must be neither too small nor too big, and the ideal is a wing length of over 320 mm. If we take one (or even two, from the bigger broods), it is placed in a cardboard box lined with grass and moss, and the others are returned to the nest. The female, and sometimes the male, will be overhead or perched in trees, but as we walk away they return, and the female will often be back on the nest by the time we reach our vehicle. That evening Tim and Brittany set off on the drive to Stafford, and so on to Poole.

I returned to Poole Harbour in early August, on an absolutely sweltering sunny day, which I could see had been the norm, for the grass fields had gone yellow. We caught up the bigger young from their cages to fit the tiny tail transmitters, but others were too small, and I was worried that some looked unfit. The next morning we were at the site at 4.30 am to release the first five young. Three flew out before long and one of them even returned to the feeding platform to eat fish by 7 am, while another briefly chased a little egret along the edge of the salt marsh. They quickly grew in confidence on the wing, and the second release was planned for the following week. One of the small ones then started to have spasms that were unfamiliar to the project team, and died. We immediately sought veterinary advice. Jamie Craig and veterinarian Emily Boyes of the Cotswold Wildlife Park were magnificent in advising and helping the team, and trying to save the young. In five days we lost five young. Expert advice suggested that this could have been brought on by the long period of extremely hot weather, which can result in birds failing to eat enough food; another thought was that there was a thiamine deficiency in the fish, so supplements were used. Tim researched thiamine and we learned that sea fish in some areas of Europe are thiamine deficient, which could impact predators of fish. The deaths reminded me of the first year of the Rutland Water project, when I was asked to try to save some runts from nests in Scotland: they subsequently died of salmonella infection. Just as it was for me then, this distressing event was a bitter blow to the team. Later, another young, which had been flying free, got entangled in waterweed and drowned. In late August and September it was a depleted cohort that set off for Africa.

On 17 October the whole team assembled at Poole to examine the results of the past summer and to come up with plans to make certain that this would not happen again. Fortunately, another landowner offered us a much better location for releasing the young ospreys, where we could have continual day-and-night surveillance of the cages, with the provision of electricity and a house where we could set up all the equipment and provide better working conditions for the team. We redesigned the cages and put them in units of two, so that fresh air could blow through much more easily in hot weather. We arranged a much better supply of fish, which would be collected daily from a trout farm, ensuring that it would be absolutely fresh; and the team would try to find a fisherman who would catch fresh grey mullet. We would also only collect large young in Scotland which were over 320 mm wing length, even though this might require two journeys to accommodate the differing ages of the young. We also decided that Brittany should be supported by two full-time assistants, as well as volunteers, during the crucial rearing and releasing stages. We were determined that 2019 would not experience the problems of the past summer but also knew that, when working with wildlife, you can never be certain. From my many decades of studying breeding ospreys in Scotland, I know that in some years a high number of chicks can die in their nests due to heavy rains, or because their male parents have failed to catch enough fish.

Paul, Jason and Brittany had carried out a lot of the preparatory work by the time Tim and I visited the new area on 20 March, so the whole team could plan the new season. I was very impressed with the new location: I could see that everything was under control and that it would prove to be a good move. We walked all over the area and decided on the best places for the release cages, the feeder nests and an array of T-perches. We were worried about an electricity power line running overhead, so Paul said he would arrange to get the company to install extra insulators on the most worrying poles. I thought everything was in order and that we could start planning to bring down a dozen young in July. Then the excitements of the summer were underway.

Blue CJ7 returned and started repairing several nests but, sadly, no migrating male osprey found her and settled down. Fortunately she kept looking, and in May one of the ringed young male ospreys we released in 2017, Blue LS7, arrived at Poole Harbour and joined her. Immediately they were visiting nests together, building them up, the male catching fish for her. As a pair, it was looking very good for the future; it was too late for this season.

We started our osprey ringing programme in 2019 on 7 July with a trip to North Sutherland and Caithness; the first nest had three superb young but it was too early to collect chicks, while the next three pairs had failed or had very small chicks. I had decided to collect the young for Poole Harbour slightly later this year, so it was not until 10 July that Tim arrived with the van. By the end of that day we had collected three excellent young. On 12 July we had a big day in the Moray forests, but found that several nests had failed and that others had very late young, so although we collected young at nests nearer my home, we only had nine young ready for the journey south. On the last day Tim drove north to Caithness and collected two big young from the

The Rutland Water bred female Blue CJ7, left, with her mate, a translocated Scottish two-year-old male photographed in a nest at Poole Harbour. (*Photo from Poole Harbour nest camera*)

first nest we had visited for ringing. Those chicks brought the number to 11, and he and Ian Perks drove them to Poole overnight, as usual.

Everything went like clockwork at the new location; Brittany, with Lucy Allen and Olivia Cooper, reared the young ospreys on freshly cut-up trout, so when we arrived for the transmitter tagging and checking day on 30 July, the young ospreys were looking really great. We caught up the first two in one cage and I was delighted at their excellent condition: feathers immaculate and a really good amount of fat already laid down on the sides of the bodies as fuel for the long migration journey. It then rained, and we had to wait until several hours after it stopped, so that the young could dry their feathers. In the afternoon we completed the necessary work and were able to report that we had 11 very fit young, ready to be released. The release date was held back until Saturday 3 August, because the electricity board were carrying out the insulation work on the Friday.

As a result I did not see the actual releases, but Brittany and Tim told me that it went very well. The team, though, had to learn to deal with a new situation, because once the ospreys were flying, they found roosting sites hidden within very large conifers nearby. This meant that the team could pick up the radio signals but couldn't see the birds. Fortunately, the excellent CCTV cameras, which also scanned the feeding platforms, allowed Brittany, Liv or Lucy to read the ring numbers and confirm each individual's presence. The team had one real disappointment: just days before the 2019 young departed on migration, one of them was predated by a fox. The post-mortem confirmed that it was a predator kill but also noted the bird was 'obese'; it is amazing how much fat ospreys lay down for migration – it can be up to 40 per cent of their bodyweight – so that word 'obese' simply meant that it was ready to migrate.

In 2020, Blue CJ7 returned and took ownership of the main nest. Then ensued a tense wait for her mate, Blue LS7, but he never returned. The spring migration was one of the worst we had recorded for ospreys – the long period of high pressure over Britain resulted in strong easterlies and poor weather over Iberia, and like other ospreys he was probably swept out to sea when trying to cross the Straits of Gibraltar or the Bay of Biscay and died out in the Atlantic. Then the Covid-19 pandemic resulted in us failing to collect a cohort of young to release, so it was a year of disappointment and delay. Hopefully Blue CJ7 will make it home in spring 2021 and a new male will join her, as the forerunners of returning the mullet hawk back where it belongs.

## White stork

As a birder, I saw the occasional lost-on-migration white stork in Scotland, usually in early spring, and I had read about the pair that nested in 1680 on St Giles' Cathedral in Edinburgh. But it was not until 1984, when I first visited East Germany, that I got to know this magnificent big white bird with its black wing markings and red bill and legs. I saw my first white storks in a huge nest on top of a magnificent oak tree in farmland close to the Polish border. My guide told me that he'd known them for many years, and that the tree was specially protected as a nature monument, so they were safe. I saw other pairs on that short visit, usually in or around villages, for local people love white storks. The following year, in northeast Poland, I saw them frequently nesting or foraging in newly cut hay meadows, and there was a pair on a chimney stack beside the Hunters' Lodge at Bialowieza, where we stayed. On 1 June I opened a trap door in the roof and could photograph the storks close up, as tree sparrows flew back and forth to their nests in the base of the stork nest. Years later I found Spanish sparrows breeding within stork nests in Spain. The next day it was heavy rain and the storks were sodden and bedraggled, and a lot less photogenic. Nothing does 'bedraggled' quite as well as a stork.

In June 1990, when we were collecting red kites in Scania, my Swedish colleagues Per-Olov and Ulf took me to see the Swedish White Stork Reintroduction Programme as we passed by. It had been started in 1989 by the Swedish Society for Nature Conservation in Skåne and the Scania Ornithological Society. The white stork had once been abundant in southernmost Sweden, with as many as 5,000 pairs during the 18th and early 19th centuries. By 1917 there were only 34 pairs, and the last pair bred in 1954. My colleagues introduced me to Karin Areskough, and as we sat outside in the sun with a coffee she told me about the project and we talked about storks. There were four pairs breeding outside the enclosure but three of them had failed because of bad weather; interestingly, a wild Polish-ringed stork had arrived to join the group. The following summer, after visiting the red kite nest and collecting two young near the stork farm, we called by and again had coffee with Karin. She told us that they had 16 young that year, but natural food was scarce because of the hot summer. As we talked, the storks were busy flying back and forth. Although I was working on a red kite reintroduction I couldn't help thinking it was about time we tried the white stork in the UK. Since then, the Swedish project has progressed and adapted: their goal is to reintroduce the white stork as a natural part of the Swedish fauna, which will involve a migratory population of 150 free-breeding pairs.

My visits to osprey projects in Portugal and Spain showed that white stork populations were burgeoning, with large numbers of nests now festooning major electricity pylons. I remember seeing about 100 nests in one area of pylons on a car journey to Lisbon. In the Basque country, Aitor Galarza, founder of the osprey project there, had carried out a small and successful reintroduction around the Urdabai Biosphere Reserve. In May 2017 Aitor took Brian Etheridge and me to meet an old farmer friend of his in the hills near Santander. It was amazing: he had 27 pairs of white storks breeding on poles of different heights in his garden. There were chicks of all ages and, with a short ladder, I could climb up and look in the nests. It had all started from one injured stork which he had cared for, and now they were migratory. They were definitely the best garden ornaments I'd ever seen.

White stork was on my list of potential projects, but it was Derek Gow and Charlie Burrell who pushed the idea through. Derek and his assistant produced a very detailed feasibility study on white storks, and Charlie offered a first release site at the Knepp estate in Sussex. I was delighted to provide an introduction to their document and to support the application to proceed. I said that there was no doubt in my mind that the white stork should again be a regular breeder in the United Kingdom and that this required dedicated translocation work, rather than waiting for the occasional pair of visiting storks to breed successfully, especially when most of the migrant storks reaching our shores are not full adults capable of breeding. The project was therefore the most certain way of restoring the species. We had no choice but to be bold and proactive, even though some of our colleagues always wrestle with perceived

An amazing white stork colony in a farmer's garden near Santander, Spain, in May 2017.

difficulties over ecology and history. They say, 'Give us your proof that this species was breeding in Britain in the past' – to which I would say, 'Give us the reason why this species, once widespread, was absent from Britain in the past.' So often they seem to forget the incredible impact of humans on larger species, especially those that were edible, in times when our ancestors were regularly hungry and desperate. If we want to be counted as responsible conservationists by future generations, we simply have to do our utmost to restore large-scale habitats and species. White stork is in the front line of those endeavours; it will be visible and known to the British public, rather than an obscure species known only to naturalists, and it will demonstrate that there is another way. Derek and his team, then, had my complete support in what I regarded as an imaginative, bold and important step in rewilding the UK.

'The Feasibility Report for the Reintroduction of the White Stork (*Ciconia ciconia*) to England' was submitted in 2016 by Derek Gow, Róisín Campbell-Palmer, Coral Edgcumbe, Tim Mackrill, Simon Girling, Helen Meech, Charlie Burrell and myself. The report assessed the potential for re-establishing a free living, breeding population of European white storks in England. Natural England and the Department for Environment, Food and Rural Affairs (DEFRA) confirmed that this species is a regular vagrant which has attempted to breed in the British Isles and that, therefore, no licence was required for a reintroduction of the species under the Wildlife and Countryside Act (1981). The Knepp estate was an inspired choice as partner, not only because the project fitted in so well with the rewilding ethos of Charlie and Issy Burrell, but because the modern Sussex town of Storrington is not far distant; in 1185 it went by the name of 'Storca-tun', old English for 'homestead with storks'. The report is available on the White Stork Project website.

Derek and Róisín soon used their contacts to start importing abandoned and rescued white storks from Poland. The birds were kept in big pens at Knepp and at other sites in Sussex, with some taken to the Cotswold Wildlife Park and Gardens, where the director, Jamie Craig, has established a successful breeding group. This provided six young in 2018 and 20 in 2019 for release into the wild. Tim Mackrill took on the role of scientific adviser

A pair of white storks breeding at Knepp Castle Rewilding Project in Sussex. (*Photo by White Stork Project*)

to the project, while a new partner is the Durrell Wildlife Trust. The breakthrough came in 2020, when the storks returned in the spring and two pairs built great nests on the crowns of large oak trees at Knepp – one pair reared three young and the other a single. It was splendid to see the full-grown flying young at their nests when I visited with my family on 2 August. Charlie and Issy were thrilled at the success as were hundreds of visitors to Knepp and thousands of followers on social media. Jamie Craig and the team at the Cotswold Wildlife Park reared another 19 young which were released at Knepp, and satellite tracking showed the storks' migration again to Iberia and even on to Morocco. It's a landmark to see them return and breed in Sussex, and to create a new English breeding population is the next goal. Just imagine the thrill in the future of seeing flocks of white storks thermalling up into late summer skies to cross the English Channel from the Isle of Wight, Beachy Head or the white cliffs of Dover.

## Restoring sea eagles to the Isle of Wight

For many years, some of us dreamed that the recovery of breeding white-tailed eagles in Scotland would one day be extended into England and Wales. As mentioned in an earlier chapter, an attempt had been made in East Anglia but had come to nothing. For me, thoughts about white-tailed eagles would often bring back memories of the Isle of Wight, because when I was a young schoolboy birdwatcher living on the Hampshire coast the island had been one of my regular haunts. The old natural history books on Hampshire and the Isle of Wight refer to sea eagles visiting the coasts in winter from the continent, and of course there was the famous last breeding site on Culver Cliff.

Each year in Scotland, as the new population continued to grow after hesitant success in the early years, I recommended, as a member of the UK Steering Group on white-tailed eagles, that our population was big enough for Scottish nests to supply young for reintroductions in the UK. It would be easier than translocating them from Norway or some other European country.

In November 2015 I gave a lecture at a mammals workshop in Winchester, and spent the night at Jonathan Spencer's home nearby. My talk had been about wolf, lynx and brown bear, and among the chat that evening we talked about bringing back sea eagles. I thought the New Forest might be a good place, along with the Solent and parts of Dorset, but Jonathan, then the Environment Manager for Forestry Commission England, said they might have some suitable woods on the Isle of Wight. Later I checked the Purbeck Army Ranges but was not impressed by the numbers of tourists on the Jurassic Coast trails, and food seemed scarce. In May 2017 Tim and I were at Poole, talking with the Birds of Poole Harbour team about the upcoming osprey project, so we made a visit to the Isle of Wight. There we met Leanne Sargent, senior ecologist with Forestry Commission England at the Lyndhurst office, and her colleague Jay Doyle. We had a very productive day and a half being shown round all the Forestry Commission woods on the Isle of Wight and taking the opportunity to look at as many places as possible to get a real feel of the island in the present day, rather than as it was in my memories from the 1950s.

A few weeks later I gathered my thoughts into a short note so that we could continue our discussions. It stated that the aim should be to restore the species to southern England so as to establish links with the emerging Irish, French and Dutch populations. The Isle of Wight had several good hacking and release sites and many areas of quiet woodland, especially in coastal landslips, to provide peaceful loafing areas. There was a good potential food supply on and surrounding the Isle of Wight, in the Solent and in neighbouring Hampshire, throughout the year. There were good potential nesting sites, mainly in trees, but also in cliff locations. There was room for approximately six breeding pairs on the Isle of Wight and excellent breeding sites in Hampshire and Dorset and, subsequently, in many areas of southern England. There were excellent locations to create 'carrion feeding' sites for future public viewing. We had a feeling that there was enthusiasm for a project within the wildlife community, tourism interests and probably from the general public. We needed to talk with sheep farmers, but farming on the Isle of Wight and south Hampshire was really different to that in northwest Scotland.

The donor young could come from Scotland, which would avoid the complications of air travel from foreign countries. Once we'd decided to carry out the project, we needed to identify partners and agree locations, and talk with all potential interests. We then needed to prepare a project plan and a feasibility report, seek and obtain funds for the project and apply for licences from SNH and Natural England. We were going to be busy, but Tim, with his recent doctorate, was ideally placed to start bringing together a fully comprehensive feasibility report, while Leanne was highly capable of dealing with a whole range of important issues, like landholdings and owners, consultation and meeting protocols, as well as allocating work on the ground.

The following March Tim collected me off a plane and we drove to Lyndhurst to meet Leanne and Jay in their office, before taking the car ferry to the Isle of Wight the following morning. We had time to walk down the woodland track to the hide at Newtown Harbour, where we looked at the brent geese and the waders, and on the way back heard a chiffchaff singing in the winter trees. Jay had arranged for us to give a presentation in Newport in the afternoon, after the morning meeting of the Local Biodiversity Action Plan, which involved the main environmental stakeholders. There were about 15 people from a range of NGOs, and Andy Davis from Natural England. I gave a talk on sea eagles and then we had about an hour's discussion; it was mostly positive or neutral, but two people foresaw a problem for the shorebirds of the Solent. I suppose that some of them already had enough on their plates and this project was coming along as an extra. We caught the ferry to Portsmouth and Tim dropped me off at Watford Junction to catch the night sleeper home.

In January 2018 the Conservative government published an important document, a 25-year plan for the Environment, launched by the Prime Minister, Theresa May, and authored by Michael Gove, the Environment Minister. Included in its aims was the reintroduction of the beaver and the white-tailed eagle. This was important and very encouraging. In early April Jonathan had asked me to give a special talk, in celebration of the coming centenary of the Forestry Commission, to their conference in Exeter, about what I thought they should do in the next hundred years. It was a chance to bring in sea eagles as well as other exciting species and projects, and an excellent opportunity for a catch-up with Jonathan on our joint sea eagle project. The next day Tim and I met with the osprey team at Poole Harbour, which allowed us to talk about the eagles, for we needed to get our eagle feasibility report completed and start the licensing process. Tim came north for our Foundation's AGM on 12 June and we had a good discussion on sea eagles; I also had a call from DEFRA, as the Minister was interested in what we were doing, so on the 14th I completed an online application with Natural England for a licence to release white-tailed eagles, which was necessary because the species is listed on schedule 9 of the Wildlife and Countryside Act. The feasibility report would follow.

Tim and Leanne were keeping us up to date on progress, and we agreed to meet on the Isle of Wight on 25 and 26 September to firm up a lot of issues. We finalised a choice of the best release site, decided where the cages should be built and discussed what else needed to be done, such as creating open perches and cutting down some trees to improve sight lines. The next day Andy Page arrived with Leanne; we met with the local forester and showed him what work was needed, and we also met the local farmer, who was very helpful and went on to become a firm friend of the project. We went over to Brightstone Forest to look at Forestry Commission woods there and also had a meeting with the National Trust manager. In the afternoon we went back to Newport to discuss our ideas with the National Farmers Union in their office, meeting the local organiser, James Osman, and the branch chairman, Matt Legge. We had a good discussion and explored some interesting points of view. They were concerned about sheep, because Matt had talked with farmers near Oban, who complained about attacks on lambs. We were sure that sheep farming in the Isle of Wight was different, but we agreed that we would keep in contact, meet farmers on the island and have a special meeting with their branch.

After that meeting I decided that it would be extremely useful to visit the Netherlands to discuss with the experts there the habits and habitats of the new breeding population of white-tailed eagles. I emailed one of the country's leading raptor specialists, Paul Voskamp, an ecologist with the Limburg provincial government. I explained to Paul that we were particularly interested in the relationship between the white-tailed eagles and the farming community, and whether there had been any complaints about the birds attacking livestock. He didn't understand my question, because nothing of the sort had happened in the Netherlands. He said he would be pleased to organise a rapid field trip for us to the white-tailed eagle breeding areas and to meet members of the Dutch white-tailed eagle working group. It was arranged that Tim and I would meet Paul on the afternoon of 25 October, and over the next two days we met two key sea eagle scientists, Dirk van Straalen, in Zeeland, and Stef van Rijk, north of Amsterdam. They told us that a pair first bred in 2006, at the Oostvaardensplassen, north of Amsterdam; that the population then rose slowly, and that in 2018 there were 18 territorial pairs, of which 11 pairs bred. We learned so much about the eagles by visiting many of their main breeding areas, with nests often close to busy places, with passing barges or motorways or towns and cities nearby, including Amsterdam on the horizon.

As dusk came in, we headed inland to the Veluwe forest, where we all stayed overnight with Dr Hugh Jansman, a research ecologist with Wageningen University. Again we were able to talk over dinner about our day and about the history and behaviour of white-tailed eagles in the Netherlands, allowing us to compare it with the situation in the UK and, in particular, with the Isle of Wight and the south of England. With regard to one of our specific requests, we were informed that in the Netherlands there had been no conflict with livestock farming. We were told there were about half a million sheep in the Netherlands; many are kept in flocks on the farms, while others graze the dykes to maintain low vegetation for dyke protection. We saw good numbers of sheep as we drove across the countryside and in several areas we saw free-range hen farms for egg production. Several of the nesting sites we visited were within half a kilometre of intensively farmed land.

Paul Voskamp works for the local government and is involved in resolving wildlife conflicts with land users, including farmers. At the present time this involves wild boar, but he has never come across or been made aware of agricultural conflicts with white-tailed eagle. Following this visit, we considered that the farming situation on the Isle of Wight was more closely comparable to that of the Netherlands and adjacent countries than to western Scotland. The Irish experience, where there had been no losses of sheep to sea eagles in 10 years, was also important in trying to allay the fears of the island's sheep farmers. We recognised the apprehensions of the Isle of Wight farming community, though, and wrote to the NFU in Newport to say that we wished to maintain dialogue. If the project went ahead, we would wish them to be a key member of any steering group.

As regards disturbance to waders and wildfowl, the Dutch ornithologists thought that it was no different to that caused by, for example, peregrine falcons, and that the bird flocks simply get used to the presence of white-tailed eagles. They probably recognise when the eagles are hunting rather than moving from place to place. It was important to note that white-tailed eagles in these habitats spend much of the day perched in a tree. In fact, on one day during our stay, we saw a hunting goshawk cause more disturbance than the eagles. The general attitude of the Dutch experts was that nearly all these species evolved with white-tailed eagles as neighbours, so the species was part of their ecosystem. Additionally, many of the birds which winter around the Isle of Wight – such as waders, ducks and brent geese – will have known sea eagles in their summer breeding grounds or on migration, so will be well aware of the hunting ability of white-tailed eagles.

In mid-November, Leanne had organised three drop-in sessions on the Isle of Wight and an evening meeting with the NFU, and had arranged various display materials and questionnaires. I caught the early morning plane to Gatwick, where I bumped into my MP and had an opportunity to talk with him about the sea eagles. I hoped he got a chance to talk with his Conservative colleague who represented the Isle of Wight. Tim collected me and we were delighted that Dave Sexton, the RSPB's Mull officer and an expert on white-tailed eagles, had joined us for the Isle of Wight visit. That afternoon we explored Brading Marsh and Culver Down, although the views of the cliffs were very limited, and we also had a good talk with the National Trust ranger. In the evening we had a meeting in the YMCA Hall in Shanklin and 32 people turned out on a cold evening; the birders were very enthusiastic and we had some very good discussions with members of the public; just one couple were there to object. The next morning, the meeting was in the Ryde scout hall, and between 11 am and 1 pm about 40 people came in to learn about the project, and to discuss it with us and share their views. Sue Crutchley and Andy Davis of Natural England arrived early so we could have a good talk beforehand, and were then able to mingle with us and the local people to hear the views. Again, it was all very positive towards the project, although a couple of people had real concerns about the risk to the island's famous red squirrels. We assured them that sea eagles cannot catch red squirrels, as shown by the scientific evidence in Scotland and in Europe.

Our meeting that evening was in the hall at Cowes Yacht Haven, where nearly 100 people turned up between 6 pm and 8 pm. Again there was overwhelming support for the proposed reintroduction. A few people, again, were concerned for red squirrels, and we had a few farmers attend, some positive, some negative. On the morning of 14 November Andy Page took Dave to see the release site and we went to Ventnor to meet Steve Jones, a local ornithologist, who wanted to help with the project. The Undercliff and the great sweep west towards the Needles looked like really wonderful sea eagle country. In the afternoon James from the NFU took us to meet a couple of farmers who bred sheep, which was very useful in helping us to understand the farming activities on the island. That evening Leanne had organised our meeting with the NFU at a cricket club near Newport. It was a good turnout, with about 70 people, including farmers and a few gamekeepers. The chairman, Matt Legge, introduced us and then handed over to me to outline our proposed sea eagle reintroduction. Before I got very far, he suggested that the meeting should have a vote on our proposal. First, he asked who was in favour. As you can imagine, not a single hand went up. The next, he asked who was against the proposal. Every hand shot up. It was a bit like an evening in the Colosseum, but at least we would leave alive. We had a long and wide-ranging discussion, with Tim, Dave Sexton, Andy Page, Leanne, Jay and myself all answering questions, and by the time we came to the end, there were really just two or three people who were pushing individual points. It was interesting when, at last, we sat down at the different tables with a drink and a bite to eat and had some interesting discussions with individual farmers. The opposition was nowhere near the level of the earlier vote. Some were genuinely interested, some had ecotourism businesses on their property, while others were arable farmers or cattle breeders with no personal concerns about eagles. We said that, once we had appointed a project officer, we would always be there to answer any queries quickly and to visit farms and get to know farmers as part of our project. We travelled home on the 9 am ferry next day, on a flat calm sea to Portsmouth.

During the winter we ran an online poll on the project, answered many queries, received important letters of support from Denmark, the Netherlands, Ireland, Spain and the RSPB, as well as the eminent ornithologist, Professor Ian Newton, and finalised the feasibility report for Natural England and DEFRA. In Scotland I outlined our plans to the winter meeting of the Highland Raptor Study Group in Inverness on 18 February, because these were the observers most likely to visit white-tailed eagle nests during the breeding season. Then, on 5 April, we received notification from Natural England that they had granted a licence, subject to some additional conditions that required further work; and so, later that month I applied for a licence from Scottish Natural Heritage to collect and translocate up to 12 white-tailed eagles per year for five years. In early May the BBC's *Countryfile* filmed a piece with me about the sea eagle project, and the press officer for Forestry England (the Forestry Commission's new name) helped by dealing with the deluge of media enquiries. In May the necessary licences were issued by Natural England and Scottish Natural Heritage and we were ready to start. We were very grateful to them for the professional and speedy way in which they dealt with our requests.

On 14 May Tim and I were in Lyndhurst to join Leanne and Andy to interview six people for the post of sea eagle project officer on the Isle of Wight; there were very good candidates and, after much discussion, we appointed Stephen Egerton-Read from the Isle of Wight, a local young man who was working for the Hampshire and Isle of Wight Wildlife Trust. In mid-May I had a very useful discussion in Inverness with Justin Grant, one of the most important white-tailed eagle field workers and the species coordinator for the Highland Raptor Study Group, and his colleague Lewis Pate; both of them were keen to help. I also discussed the collection of young with Kenny

Kortland of Forestry and Land Scotland and with SNH, but in the first year I was cautious about taking young from the specially monitored pairs in Argyll. In mid-June Ian Perks and I went to meet Dave Sexton in Mull and discussed opportunities there. During the day we went to a nest with Dave and his team, where a single chick was ringed; Dave told us he knew of at least two other chicks, which would be available later.

On 17 June we finished getting the pens ready at my home for holding the eagles before translocation and arranged a supply of fish and rabbits to feed them. Ian Perks and Fraser Cormack went to the island of Mull on 19 June and, with Dave Sexton, collected two really good young – a male and a female – from different nests. Anya Wicikowski arrived from Manchester to take on the task of preparing the food and feeding the young. On 20 June Ian, Fraser, Tim and I went to Wester Ross to collect a young from a big nest in a tall Scots pine, where I'd first seen breeding sea eagles 20 years previously, and then we met up with Seamus McNally from Torridon who took the three younger guys by boat to a nest on an island, where one young was collected, meaning that we returned home with two eaglets. I had decided that this first year we would translocate six birds, so with the help of permissions from a forestry company and Alison MacLennan of the RSPB on Skye, Tim, Anya and Fraser went to the Isle of Skye and collected chicks from two eyries in forests. Meanwhile I went to Burghead to get a supply of fresh fish. On 24 June Tim and Fraser set off overnight in a big van for the Isle of Wight with the precious cargo of six young white-tailed eagles.

Much work had been done on the Isle of Wight, especially on the superb hacking cages constructed in steel by Pete Campbell, with the woodwork finished by his friend Dick. When I saw them I was absolutely delighted and said they were the best eagle hacking cages I had ever seen. At the release site our friend Jason Fathers had installed a superb CCTV system which ran back to a caravan concealed in the bushes at the edge of the field. Tim and Fraser arrived the next morning, with the eaglets, to be greeted by a welcoming party from the sea eagle team, volunteers, the local farmer and Mark and Mo Constantine, who were supporting us with the project. It was a momentous day. The routine job for Steve and his volunteers was to keep the birds well fed and to maintain a record of all their behaviour, using the CCTV cameras. During this period there was a short time when we thought that one of the birds was being bullied by the largest female, and that it might have to be moved into another cage, but that was solved by putting food for the two birds through the back of the cage in two different places.

In late July Tim and I had a very nice evening with Jonathan Spencer and his wife Alison in Winchester, and were joined for supper by Paul Goriup of great bustard fame. We had a really

Young sea eagles in a nest in Wester Ross – one was collected for the Isle of Wight project (*Photo by Ian Perks*)

great evening of talking, and next day Jonathan, Tim and I went to the first meetings of the white-tailed eagle steering group in the morning and the monitoring group in the afternoon. It was very good to hear all the discussions and to get these two groups underway. We were going to release the eagles in the second week of August, until a friend told us that was the week of the famous Cowes Regatta, with thousands of yachts and people on the Solent as well as a display by the Red Arrows and massive fireworks displays. We delayed until 19 August, when we assembled in the evening on the Isle of Wight. The next morning John Chitty, the raptor specialist veterinarian, arrived from Southampton and we were ready to capture all of the eagles to weigh and measure them, to fit the satellite transmitters and for John to check their health and take blood samples for analysis. These are very, very big birds, which have to be held extremely securely and competently, so it was really good that I had Tim and Ian with me, as they have gained a lot of experience with sea eagles in Scotland.

Everything went well, and next day we were at the caravan by 4.30 am before dawn. Pete and Steve were already there, as well as the BBC *The One Show* crew and presenter Mike Dilger. Pete and Steve released the front door in the cage which held the two Mull birds, and we all watched the monitor in the caravan; finally the female walked out onto the outside perch but didn't fly off, and then the male, which had been bullied by her, flew straight past and soared out over the meadow before flying down to perch in the wood. Before too long, the big female followed, also perching in the nearby woods. A *Daily Telegraph* photographer got some good photographs from the top of our tower built in the edge of the forest. The next day we released the other four and none of them travelled any great distance, all perching in trees within about a mile. Finally it was time for us to go home and leave them in the care of Steve and the Isle of Wight team. I was thrilled to see the excitement of the local volunteers, such as John Wilmott and Jim Baldwin, who put in so much time. They so enjoyed the fact that this iconic project was happening on their island.

We've learned a lot since then. Four eagles have stayed on the Isle of Wight, mainly living in quite small areas, making occasional flights around the island and visits to the food platform near the cages. One bird – called Culver, after the ancient cliff nesting site – visited the New Forest and then in September made an amazing six-day tour of southeast England, including flying directly over the Houses of Parliament in central London. It roosted overnight on the Essex coast and then, next day, started a leisurely journey back to the Isle of Wight. Only three people reported seeing the bird on that trip. Later on, another eagle headed north and took up residence in Oxfordshire, where he was often followed by small flocks of red kites and buzzards. When he was on the ground, eating carrion, magpies would

A pair of young white-tailed eagles released on the Isle of Wight, October 2019. The female is carrying a stick. (*Photo by Nick Edwards*)

hop up to see this new bird on the block. Sadly, one of the eaglets died on the island; it had been eating a dead porpoise on the beach for three weeks, and the eagle's body was duly sent for analysis. Meanwhile radio signals stopped coming in from Culver, the bird that had flown to London, and its whereabouts are unknown.

Four of the young sea eagles survived their first winter, one of them spending it in Oxfordshire. In March they became more active and sightings in southern and eastern England included at least five young white-tailed eagles from mainland Europe. In April, two of the young flew north and spent the summer in the North York Moors where they fed on rabbits in the valleys. On the Isle of Wight, Steve proved that the eagles were starting to catch and eat mullet in the estuaries, but then on 31 May the remaining female left the island and slowly flew north to the southern shore of the Firth of Forth. We wondered if she was going back to Wester Ross or would she return; instead she summered in the rabbit-haunted valleys of the Moorfoot Hills in southern Scotland. In late August she slowly flew back to the Isle of Wight, in fact straight back to the release location. She joined her partner of winter 2019/20 and they became efficient catchers of mullet, bream, bass and cuttlefish in the rich waters around the island.

Seven more young sea eagles were released in late July and this time we released them at an age closer to normal fledging, which resulted in them becoming very well hefted to the release site meadow and surrounding woods. From the start, Steve had been putting out plenty of dead rabbits and fish – because of Covid-19 it was a solitary task. Like last autumn the young birds fitted very well into the countryside of the island; one sadly died after striking an electricity power line and by the end of the year, four were still on the island with singles in North Cornwall and the Devon / Somerset border. The older eagles, which summered in Yorkshire, moved south to winter in east Lincolnshire and northwest Norfolk. The two-year old pair on the island was usually together and the male in particular gave splendid displays of fishing out in the English Channel with flocks of gulls. He even ate the smaller fish while in flight and flew up to 7 kilometres offshore – exciting behaviour which we did not foresee and which is very encouraging for the future. These are early days, of course, but it's been amazing what a tremendously positive response there has been to this exciting project and we are very encouraged that the eagles have located excellent resources of fish and learned how to find rabbit warrens in the English countryside. This rewilding story is only just beginning but is already looking encouraging.

Appendix

# THE FOUNDATION AND OTHER ORGANISATIONS

## Roy Dennis Wildlife Foundation (RDWF)

The Foundation was established in Scotland in 1995 as the Highland Foundation for Wildlife (HFW). It is a non-membership charitable trust dedicated to wildlife conservation and research, with a special emphasis on species recovery projects and the restoration of natural ecosystems. Dr Tim Mackrill joined me to work on projects in England and elsewhere in 2016, and a year later the name was formally changed to the Roy Dennis Wildlife Foundation. See our website: www.roydennis.org

**ACPBS** – Advisory Committee on the Protection of Birds in Scotland
**BTO** – British Trust for Ornithology
**CCW** – Countryside Council for Wales
**CITES** – Convention on International Trade in Endangered Species of Wild Fauna and Flora
**DEFRA** – Department for Environment, Food and Rural Affairs
**ERW** – Eagle Reintroduction Wales

**FAPAS** – Fund for the Protection of Nature (in Asturias, Spain)
**HFW** – Highland Foundation for Wildlife (now RDWF)
**IUCN** – International Union for Conservation of Nature
**LPO** – Ligue pour la Protection des Oiseaux (French Birdlife)
**MoD** – Ministry of Defence
**NATO** – North Atlantic Treaty Organization
**NCC** – Nature Conservancy Council
**NFU** – National Farmers' Union of England and Wales
**NGO** – Non-Governmental Organisation
**NINA** – Norwegian Institute for Nature Research
**NNR** – National Nature Reserve
**RAFKOS** – RAF Kinloss Ornithological Society
**RDWF** – Roy Dennis Wildlife Foundation (formerly HFW)
**RSPB** – Royal Society for the Protection of Birds
**RZSS** – Royal Zoological Society of Scotland
**SNH** – Scottish Natural Heritage
**SOC** – Scottish Ornithologists' Club
**SSSI** – Site of Special Scientific Interest
**SWT** – Scottish Wildlife Trust
**WWF** – Worldwide Fund for Nature (formerly the World Wildlife Fund)

# ACKNOWLEDGEMENTS

The wildlife projects in this book have been joint efforts by groups of friends and colleagues, often working together at different times on different species. I sincerely thank them all for decades-worth of encouragement and support, often of a practical nature and in challenging conditions. Mentors like John Everett, Peter Davis, George Waterston, Dick Fursman, John Morton Boyd, Hugh Boyd and Derek Ratcliffe encouraged me in the early years, while Ian Newton and Mick Marquiss were always there to give me scientific advice and encouragement. During the goldeneye and the early sea eagle years, my assistants Roger Broad and, later, Colin Crooke, both working from the RSPB Highland Office, were great colleagues, as has been – in more recent times – Tim Mackrill, my colleague in the Roy Dennis Wildlife Foundation. Since the mid-1970s, Charles Fraser, Nigel Graham, John Grant, Bill Templeton, Jamie Whittle, Lucy Lister-Kaye, John Nicolson and Frank Law have been trustees of the Highland Foundation for Wildlife, subsequently renamed the Roy Dennis Wildlife Foundation. I thank them very much for their encouragement of, and enthusiasm for, our ideas; and also the many who funded our conservation work. I also thank staff in the old Nature Conservancy Council, Scottish Natural Heritage, Natural England and government departments dealing with licensing and animal health, for valued advice and licensing arrangements.

Many birders and naturalists have helped with our work in this country and abroad, and I thank them all. My life and work have been greatly helped by the kindnesses shown by landowners, factors, gamekeepers, farmers, crofters, foresters and stalkers who have given me access for my wildlife work. In order to protect the locations of rare bird nests, their names remain confidential, but I am grateful to them all for their support and friendship, often over many years. I feel privileged to have worked in such wonderful parts of the world. The Forestry Commission staff have been equally supportive.

With regard to the individual projects, I am indebted to many friends and colleagues over the decades:

**Sea eagles at Fair Isle:** George Waterston, Johan Willgohs, Tony Mainwood, Dennis Coutts and the Fair Islanders of the 1960s. **Rum and Loch Maree reintroductions and breeding attempts:** John Morton Boyd, Ian Newton, Martin Ball, John Love, Doug Weir, Roger Broad, Colin Crooke, Harald Misund, Dave Sexton, Mike Madders, Kate Nellist, Ken Crane, Dougie Russell, Stewart Keenan, Greg Mudge, Steve Rooke, Kevin Duffy and Alison MacLennan.

**Goldeneye project:** George Waterston, Pat Sandeman, Dick Fursman, Hilary Dow, John Parslow, Stuart Taylor, Roger Broad, Dave Pierce, Dave Pullan, Doug Weir, Colin Crooke, John Lister-Kaye, Tony Hinde, Bob Swann. **Peregrines:** Derek Ratcliffe, Christian Saar, Vero Wynne-Edwards, Tony Colling, Dick Balharry, Lorcan O'Toole and Jimmy Maclean.

**Red kite reintroduction:** Magnus Sylvén, Nils Kjellén, Johnny Karlsson, Per-Olov Andersson, Ulf Sandnes, Steve Rooke and the RAF Nimrod crews and Kinloss station commanders, Colin Crooke, Ian Newton. Peter Davis, Richard Porter, Roger Lovegrove, Leo Batten, Mike Pienkowski, Ian Evans, Frank Hamilton, Bobby and Kitty MacDuff-Duncan, Peter Fennel, John Easton, Dee Doody, Andy Knight, Duncan Orr-Ewing, Lorcan O'Toole and Brian Etheridge.

**Ospreys to Rutland Water:** Tim Appleton, Martyn Aspinall, Stephen Bolt, Colin Crooke, Phil Grice, Ian Carter, Helen McIntyre, Hugh Dixon, Mick Marquiss, Mark Martell, Jo Hayes, Bob Moncrieff, Andy Douse, Tim Mackrill, Paul Stammers, Andy Brown, Barrie and Tricia Galpin, Lloyd Park and John Wright. **Ospreys in Wales:** Roger Lovegrove, Graham Williams, Blaydon Holt, Tony Cross, Clive Faulkner, Steve Watson, Emyr Evans and Janine Pannett.

**Red squirrels:** Ted Piggott, Donald and Margo Maxwell MacDonald, Jane Rice, Alastair Macdonald, Lesley Cranna, Tamara Lawton, Ben Ross, Ian Collier, Juliet Robinson, Derek Gow, Craig Shuttleworth, Frank Law, Bill Cuthbert, Stephen Corcoran, Jane Harley, Gaby Bongard, Anna Meredith, John and Anne Lychet, Ron Macdonald, Becky Priestley, Fiona Newcombe, Innes MacNeill, David Clark, Jonny Shaw, Andrew Sutherland, Alistair Sutherland, Paul Lister, James and Carol Hall.

**Beavers:** Derek Gow, Karl-Erik Jonsson, Don MacCaskill, Vilmar Dykestra, Bart Nolet, Sabine Hille, Mark Harthorn, Dietrich Heidecke, Annett Schumacher, Gerhard Schwab, Paul Ramsay, Duncan Halley, Magnus Magnusson, John Lister-Kaye, Roger Crofts, Goran Hartmann, Olivier Rubbers, Achille

Verschoren, Kenny Taylor, Niall Benvie, Hugh Chalmers, Allan Bantick, Martin Gaywood, Roger Crofts, Colin Galbraith, John McAllister, Dick Balharry, David Windmill, Iain Valentine, Simon Milne, Simon Jones and Roo Campbell.

**Large carnivores and herbivores:** Christoph and Barbara Promberger, Wolf Schröder, Sigrid Rausing, Reidar Andersson, Urs and Christine Breitenmoser, Magnus Sylvén, Vin Fleming, David Hetherington, Reay Clarke and Luis Palma.

**Eagles in Ireland:** Lorcan O'Toole, Allan Mee, Jeff Watson, Jim Haine, John Marsh, Ronan Hannigan, Kevin Collins, Hugh Insley, Colin Crooke, Kevin Lawlor, John Easton, Marc Ruddock, Justin Grant, Doug Mainland, Derek Spenser, John Braine, Richard Nairn, Charlie Haughey, Tony Whilde, Duncan Halley, Torgeir Nygård and Alv Ottar Folkestad.

**Osprey projects in Europe:** Luis Palma, Pedro Beja, Rita Alcazar, Andreia Dias, Miguel Ferrer, Eva Casado, Juan Rubio, José Sayago, Pertti Saurola, Rolf Wahl, Mikael Hake, Daniel Schmidt, Manuel Pomarol, Giampiero Sammuri, Alessandro Troisi, Flavio Monti, Jean-Claude Thibault, Jean Marie Dominici, Andrea Sforzi, Pietro Giovacacchini, Aitor Galarza, Ben Ross, Tristan Bradley, Gabriella Tamasi, Mike Dilger, Ian Perks, Fraser Cormack, Brian Etheridge, Wendy Strahm, Denis Landenbergue, Oliver Biber, Michel Beaud, Luc Hoffman, Martin Herbach, Dave Anderson, Eladio Fernandez-Galiano, Paul Lesclaux and Florent Lagarde.

**Cranes:** George Archibald, John Buxton, Zoe Taylor, Wolfgang Mewes, Mikael Hake, Oliver Krone, Justin Prigmore, Thomas MacDonell, David Hetherington and Hywel Maggs. **Golden**

**eagles at Kielder:** Brian Little, Bill Birleton, Ian Newton and Graham Gill. **Great bustard:** Gordon Barnes, Aylmer Tryon, Max Dornbusch, David Waters and Paul Goriup. **Eagle owl:** Colin Crooke, Fergus Beeley, Paul Voskamp, Adrian Aebischer and Tony Crease.

**Sea eagles to Suffolk and Europe:** Tim Appleton, Andy Brown, Julian Roughton, Alan Miller, Richard Saunders, Frans Vera, Phil Grice, Peter Newbury, Adam Gretton, Torsten Langgemach, Adam Rowlands, Tadeusz Mizera, Duncan McNiven. Mick Marquiss, Steve Watson, Nick Pinder, Duncan Halley, Torgeir Nygård, Doriano Pando, John Cortes, Stephen Warr, Miguel Ferrer, Jean-Marc Thiollay, Rolf Wahl, LC Kvaal and Juan José Sánchez Artés.

**Ospreys on the south coast of England:** Colin Tubbs, Mark and Mo Constantine, Tim Mackrill, Fiona Macphie, Barry Duffin, Alan Poole, Barry Collins, Pete Potts, Manuel Hinge, Bob Chapman, Andy Page, Mark Singleton, Dante Munns, Jason Fathers, John Wilmott, Ben Ross, Paul Morton, Emily Joachím, Brittany Maxted, Jamie Craig, Emily Boyes, Lucy Allen, Olivia Cooper, Fraser Cormack, Ian Perks and Brian Etheridge.

**White storks:** Karin Areskough, Derek Gow, Charlie Burrell, Isabella Tree, Penny Green, Lucy Groves, Tim Mackrill, Róisín Campbell-Palmer, Jamie Craig, Emily Boyes and Reggie Heyworth.

**Sea eagles to the Isle of Wight:** Jonathan Spencer, Tim Mackrill, Leanne Sargeant, Jay Doyle, Ian Newton, Andy Page, James Osman, Matt Legge, Paul Voskamp, Dirk van Straalen, Stef van Rijk, Hugh Jansman, Dave Sexton, Su Crutchley, Nikki Hiorns,

Andy Davis, Richard Saunders, Giles Wagstaff, Ben Goldsmith, Stephen Egerton-Read, Justin Grant, Lewis Pate, Kenny Kortland, Fraser Cormack, Ian Perks, Anya Wicikowski, Seamus McNally, Dave Sexton, Alison MacLennan, Pete Campbell, Steve Jones, John Chitty, John Wilmott and Jim Baldwin.

I record my sincere thanks to Tim Mackrill for reading and commenting on an early draft and to my wife, Moira, for improving the grammar and structure. James Macdonald Lockhart agreed to be my agent and has offered me great support and encouragement in writing this book and more. Myles Archibald, commissioning editor at HarperCollins, responded enthusiastically and suggested a wise restructuring of the chapters, while Hazel Eriksson, my editor, has guided me through the production of the book. My sincere thanks to you all. My first wife, Marina, and my three older children, Rona, Gavin and Roddy, were involved in many ways in the earlier projects, while Moira and our young daughter Phoebe have helped with the later ones. Without the support of my family and colleagues, my life spent restoring the wild would simply not have been possible.

# INDEX

Page references in *italics* indicate images.
RD indicates Roy Dennis.